Practical Analog and RF Electronics

Practical Analog and RF Electronics

Daniel B. Talbot

CRC Press
Taylor & Francis Group
Boca Raton London New York

CRC Press is an imprint of the
Taylor & Francis Group, an **informa** business

First edition published 2021
by CRC Press
6000 Broken Sound Parkway NW, Suite 300, Boca Raton, FL 33487-2742

and by CRC Press
2 Park Square, Milton Park, Abingdon, Oxon, OX14 4RN

First issued in paperback 2022

Library of Congress Cataloging-in-Publication Data

Names: Talbot, Daniel (Daniel B.), author.
Title: Practical analog and RF electronics / Daniel B. Talbot.
Description: First edition. | Boca Raton : CRC Press, 2021. | Includes
bibliographical references and index.
Identifiers: LCCN 2020020919 (print) | LCCN 2020020920 (ebook) | ISBN
9780367542917 (hbk) | ISBN 9781003088547 (ebk)
Subjects: LCSH: Radio circuits. | Analog electronic systems.
Classification: LCC TK6560 .T335 2020 (print) | LCC TK6560 (ebook) | DDC
621.3841/2--dc23
LC record available at https://lccn.loc.gov/2020020919
LC ebook record available at https://lccn.loc.gov/2020020920

ISBN: 978-0-367-54291-7 (hbk)
ISBN: 978-0-367-54294-8 (pbk)
ISBN: 978-1-003-08854-7 (ebk)

DOI: 10.1201/9781003088547

This book is dedicated to the engineer, who is truly indispensable to civilization.

Contents

Preface

It is the author's intention to present practical yet simple approaches to solving problems in filters, analog circuitry, and radio-frequency (RF) designs, and to inform the reader of approaches that are generally not covered in other texts: voltage gain from a passive network without using transformers, matching an impedance using a single component but without using a transformer, fixing a surprise voltage gain in a single diode detector, realizing an infinitely (yes, infinitely) deep notch from a passive network with losses (finite Q), demodulating an amplitude-modulated (AM) signal using a non-conventional local oscillator waveform to demodulate said signal without carrier or twice-carrier or sum-and-difference or image frequency residue in the result, and realizing a passive group delay equalizer that allows for lossy components. The standard Darlington bipolar junction transistor (BJT) gain block is analyzed and simulated. These are just some of the issues addressed in this book.

We will begin this book with a discussion of operational amplifiers (op-amps, pardon if we start simple) and then discuss passive networks commonly used with those op-amps. We will also discuss the usefulness of the generalized Miller effect in quickly calculating closed-loop op-amp-based behavior. We will also discuss short-circuit transfer impedance, and amplification in the current domain rather than the voltage domain as a way to circumvent the Miller effect (to a large degree), and the property of reciprocity in networks and how it affects design.

For extreme cases of signal processing where the source is a current source or nearly so, to achieve the highest signal to noise (SNR) ratio, we will discuss the transimpedance amplifier.

This book is meant to be a handbook (or a supplemental textbook) for students and practitioners in the design of analog and RF circuitry with primary emphasis on practical albeit sometimes unorthodox circuit realizations.

We will present plenty of techniques that a seasoned professional can adapt to a special design requirement, such as the notch networks that have infinite depth but with lossy components, and the unique 2nd order passive delay equalizer, as well as a simple method of accomplishing capacitance neutralization using a negative capacitor.

Simple single-component impedance matching techniques that work in most narrow bandwidth bandpass filter situations are described in a step-by-step fashion. This method is accurate, quick, and simple.

The use of the secant waveform to demodulate a double-sideband suppressed-carrier (DSBSC) or non-suppressed-carrier signal is demonstrated to be superior to conventional synchronous demodulation due to the minimized carrier and twice-carrier frequency residue, but also the suppression of certain image frequency signals. The secant waveform truncated at 15 to 1 crest factor can be generated with a look-up table (e.g., programmable read-only memory – PROM) driving a digital-to-analog (D/A) converter quite simply (and can build in predistortion of the secant waveform to cancel distortion in the mixer), and when used as a synchronous local oscillator to demodulate a carrier at RF or IF (intermediate frequency), the result is nearly perfect recovery of the modulation envelope without any glaring artifacts.

We will present the important formulae and text to solve for the locking range of the injection locking phenomenon in oscillators. Phase-locked loops (PLL) design will be presented including a short computer program for automating some of the design. The choice of when to use a digital versus analog PLL is explained. A false locking prevention method is presented which guarantees no locking on a sideband or spur using the DC offset from a self-sweeping technique (assuming an analog phase detector).

Second and third order distortion ramifications on two-tone stimuli are presented, as is the description of intermodulation, cross-modulation, and spectral regrowth phenomena. The "one deciBel compression point" resultant two-tone intermodulation level of −27.8 deciBels is explained and used to replace the concept of third order intercept point.

Optimization is also discussed and examples are given. Again, here the emphasis is on the practical.

In the final chapter, we discuss and illustrate quadrature distortion in the waveform recovered from an AM signal impaired by asymmetric sidebands due to filter asymmetry and show how synchronous detection solves the problem. Lastly, we discuss cross-rail-interference, also caused by filter asymmetry.

A list of references (some of which are intensive) exists in the appropriate section which relates to some of the topics but not necessarily the techniques discussed in this book. Because much theory exists outside this text, it will not necessarily be repeated here, so the reader is encouraged to use the references to dig deeper. However, we will attempt to inform the reader of most of the basic theory.

Dan Talbot

About the Author

Daniel B. Talbot has been named a Life Member of the Institute of Electrical and Electronics Engineers (IEEE) for his membership of over 50 years and a member of Eta Kappa Nu, and also a Fellow of the Audio Engineering Society for his accomplishments while Chief Engineer of DBX, an audio equipment manufacturer making noise reduction products.

He was a Principal Engineer at Raytheon Missile Systems, working mainly on frequency synthesis.

He was a Research Engineer at David Sarnoff Laboratories (RCA), working on issues in color television such as synchronous demodulation of the vestigial sideband (VSB) transmission system and surface wave filter side effects and was involved in integrated circuit (IC) design (using an early 3-GHz RCA process).

He has worked on radar systems concentrating on achieving high sub clutter visibility (SCV) while at Sanders Associates (later purchased by Lockheed) and state-of-the-art frequency synthesis.

As a consultant to Harris Broadcast division he helped design the DIGIT digital FM exciter used in their frequency modulation (FM) transmitter products. He also designed frequency synthesizers for several television transmitter models to improve incidental phase modulation (ICPM) and for special international requirements.

As Chief Engineer for consumer electronics testing at LTX, he designed their first RF signal generator and color television pattern generator.

He also designed integrated circuitry for a high-speed cable TV modem using a 60-GHz Silicon-Germanium (SiGe) bipolar process.

As a consultant, he has designed state-of-the-art fiber-optic receivers for several clients.

His career as an engineer began in 1968 when he graduated from the University of Nebraska Electronic & Electrical Engineering (EE) Dept. He holds eight US patents and has published 17 papers.

His interests include classical music, opera, photography, gourmet cooking, and ballroom dancing.

Some of his more notable papers include:

(1) "N-Over-M Frequency Synthesis," *Rf Design Magazine*, September 1997.

(2) "A 200 KHz to 130 MHz Direct Frequency Synthesizer for Ultra-Low Distortion AM And FM Applications," Proceedings of Semicon Europa, Zurich, Switzerland, March 1981.

(3) "Ultra-High-Performance Amplitude and Frequency Modulation and Demodulation," *Journal of the Audio Engineering Society*, July/August 1981, pp. 492–502.

(4) "Wide-Dynamic-Range Scalar Network Analyzer (with Low-Cost Group Delay Measurement Feature)," *Microwaves & RF*, January 1997.

(5) "A Review of Non-Linear Distortion Fundamentals," paper presented to AES Convention, New York, October 1984.

(6) "A Satellite Communications, Broadcast-Quality Amplitude Compander," *Journal of the Audio Engineering Society*, October 1981.

(7) "Fiber-Optic Transmission and Professional Audio," *Journal of the Audio Engineering Society*, May 1994.

1

Operational, RF, and Current Amplifiers and Their Ubiquity

1.1 Introduction

Operational amplifiers, inverting and non-inverting mode and the reason the latter exhibits better signal-to-noise ratio (SNR), and operational transconductance amplifiers (OTAs) and class-C and F high-efficiency amplifiers are explained in this chapter. Reciprocal networks allow input and output terminals to be swapped without consequence to gain in an op-amp application, but with a possible consequence to noise.

A short-cut is presented using the Miller equation to "back-of envelope" calculate closed-loop gain. The Miller effect is useful for capacitor neutralization.

Description of the transistor as either a current conveyor (grounded base or gate) or as a transconductance amplifier (grounded emitter or source) is provided. High bias current for improving SNR is discussed, with concomitant discussion of shot and resistor noise.

This chapter also includes discussion of gyrators to synthesize inductors.

A high-frequency amplifier circuit is used that is a two-transistor Darlington with feedback. The circuit is modeled and analyzed.

The current conveyor is discussed and its advantage for both high gainbandwidth (GBW) and high dynamic range. By handling signal currents rather than voltages, one largely escapes the Miller effect.

Simulation Program with Integrated Circuit Emphasis (SPICE) calculations of the linearity of the open-loop nature of a current conveyor and closed-loop nature of an op-amp are compared.

Cascode circuits are not immune from the Miller effect in layout.

DOI: 10.1201/9781003088547-1

1.2 The Op-Amp and Its Real and Imaginary Parasitics and Compensation

The symbol for a standard op-amp is shown in Figure 1.1. Two inputs are shown; one is the non-inverting (+) and the other (-) is the inverting input. The output is the difference of the two input voltages at those ports multiplied by the gain (A) of the op-amp. As shown, there may be a shutdown input which turns the amplifier on or off, and when in the off state, the output is usually forced to become an open circuit. There are of course power supply terminals, which may be bipolar or unipolar, and when the former, they form a pair. Typical values are +15 and −15 volts or higher, but they can be as low as +5 and −5 volts or lower. In some cases, only one supply terminal (also called a rail) exists and the op-amp output can then have only one polarity (unipolar). So much for the idealized case. The practical case has limited bandwidth, slew rate, (common mode rejection ratio (CMRR), power supply rejection ratio (PSRR), output current available, input noise voltage, input noise current, supply current drain, input balance (offset), behavior over temperature, and process (e.g., silicon-germanium, metal-oxide-semiconductor (CMOS), standard silicon bipolar junction transistor (BJT)-based), input bias current, survivability to electromagnetic or gamma pulse (EMP) via a process called dielectric isolation, and the ability to handle large input common mode and or differential input voltages). One of the first commercial op-amps to exploit integrated circuit technology was the uA709, designed by Robert Widlar [1, 2]. The schematic diagram is shown in Figure 1.2.

A model of a practical op-amp is shown in Figure 1.3. It features a gain term G, an input resistance *RI*, an output resistance *RO*, and a gain-band-width product (GBW) *F*.

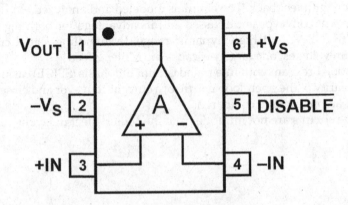

FIGURE 1.1
Op-amp with shutdown.

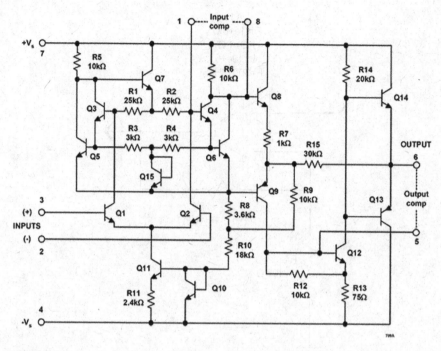

FIGURE 1.2
Schematic of ua709 op-amp, requiring external compensation capacitor.

RI=

RO=

G=

F=

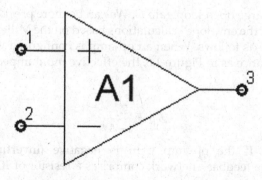

FIGURE 1.3
Practical op-amp model, open loop.

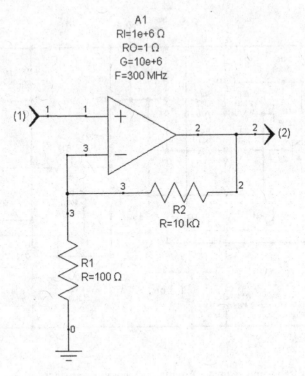

FIGURE 1.4
Closed-loop gain of 101 at zero Hz.

This op-amp can be configured per the example of Figure 1.4 to have a closed-loop gain of 101, and a 3 dB bandwidth of 300 MHz divided by 101, or about 3 MHz (see Figure 1.5). In this configuration, closed-loop gain at zero Hz is given as:

$$G_{closed} = 1 + \frac{R2}{R1} \qquad (1.1)$$

assuming very large open-loop gain G. We can be more precise by factoring in some "back of the envelope" calculations based on the Miller effect, which we will describe as follows. When an op-amp is configured with a solitary feedback impedance as in Figure 1.6, the effective input impedance is equal to:

$$\mathbb{Z}_{IN} = \frac{\mathbb{Z}_F}{1 - A} \qquad (1.2)$$

See Figure 1.7. If the op-amp gain is negative (inverting) 1000 (for example) and the feedback network comprises a resistor of 1000 Ohms and

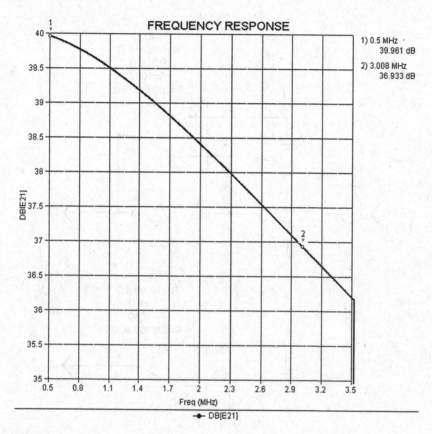

FIGURE 1.5
Closed-loop bandwidth of circuit of Figure 1.4.

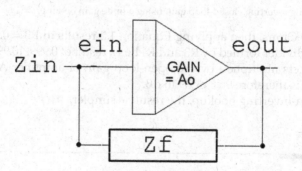

FIGURE 1.6
The Miller effect.

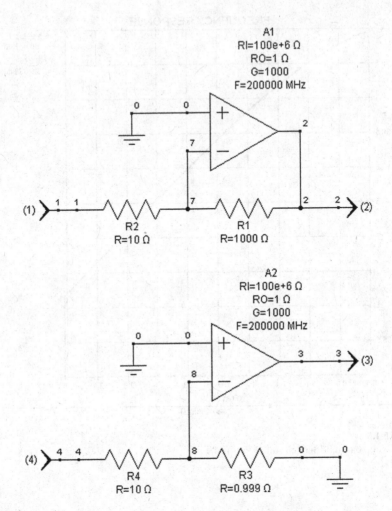

FIGURE 1.7
Aid to calculating inverting closed-loop gain using a finite gain op-amp.

R2 = R4 = 10 Ohms, then applying Equation 1.2 results in R3 = 0.999 Ohms. The voltage divider formed by R3 and R4 has a result of 0.999/10.999 = 0.0908. This result gets multiplied by the open-loop gain of amplifier A2, or 1000. The product is therefore 90.8 or 39.16 dB.

For the non-inverting hookup, the result is simpler.

1.3 Real and Imaginary Parasitics

Observe a simplified op-amp model in Figure 1.8 Transistors Q3 and Q6 are assumed to be a matched pair (easily accomplished in an integrated circuit) forming a current mirror [2, 3]. Together, they repeat the bias current from

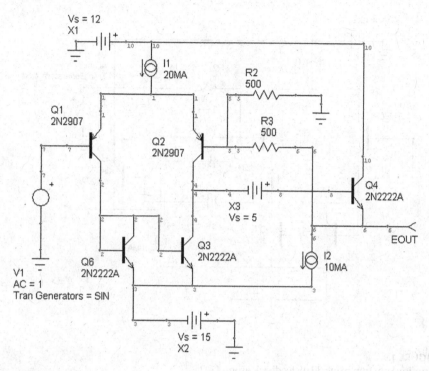

FIGURE 1.8
Primitive op-amp using discreet ideal components.

Q1 so that it matches the bias current from Q2, forming a gain stage. The real output load presented to Q2 is the load at emitter follower Q4 multiplied by the beta of Q4. The load capacitance on Q2 output is assumed negligible but Q2's capacitance between collector and base (C_{ob}) is non-negligible and forms a rolloff illustrated by Figure 1.10 and Figure 1.11 when it is 33 pF external (Figure 1.9 for schematic).

1.4 Compensation

As open-loop gains become large, poles (in the s-plane) of each subsequent gain stage internal to the op-amp represent phase lag. If the phase lag nears 180 degrees at unity gain open-loop, we need to shift the open-loop response to roll off much sooner, otherwise we have an oscillator. Integrated circuit op-amps have very large gains and would oscillate closed-loop if the phase were not compensated. In our primitive op-amp model, we added 30 pF around the base-collector of Q2. It is effectively in parallel with the feedback resistor. Although not needed in the example, a capacitor (eg 10-40 pF) is normally added from base to collector of Q3. This process is called "dominant pole stabilization."

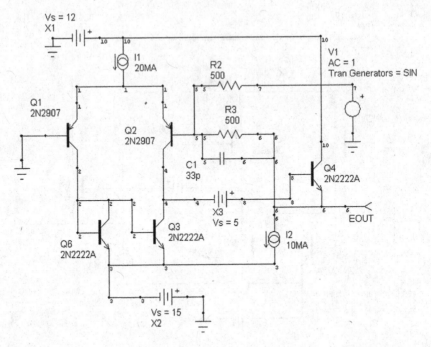

FIGURE 1.9
Primitive op-amp using 30 pF feedback around Q2.

In BJT designs a single stage can have tremendous gain. Refer to Figure 1.12, where the transconductance gain of a grounded-emitter BJT is shown to be

$$gm = \frac{qI_c}{kT} \tag{1.3}$$

where
    I_c = the collector bias current in amps
    q = the charge of an electron = -1.602×10^{-19} coulombs
    k = Boltzmann constant = 1.3806×10^{-23} m^2 kg s^{-2} K^{-1}
    T = temperature in degrees Kelvin

This reduces to

$$gm = 42I_c \tag{1.4}$$

So that at 1 mA collector bias current, the transconductance gain is 42 mmhos, which, when the collector is loaded with a 1K-Ohm load, yields a voltage gain of 42 (see Figure 1.13).

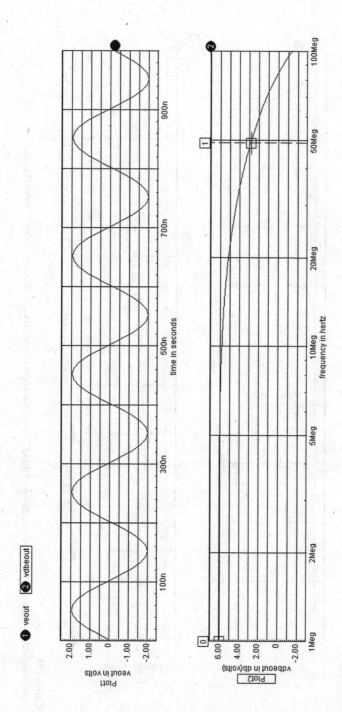

FIGURE 1.10

Primitive op-amp's gain and frequency response with Q2's "Miller parasitic capactive" term C_{ob}.

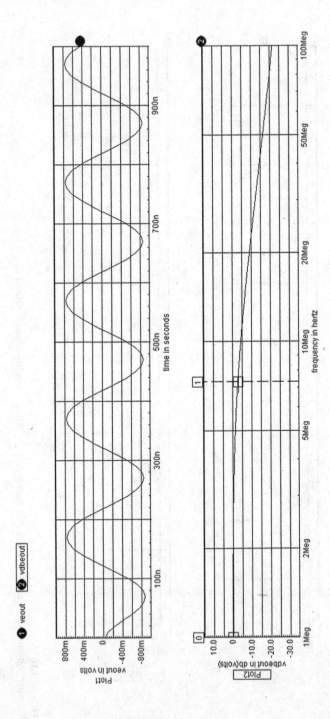

FIGURE 1.11

Primitive op-amp's gain and frequency response with Q2's parasitic capactive term C_{ob} and 33 pF added external to Q2.

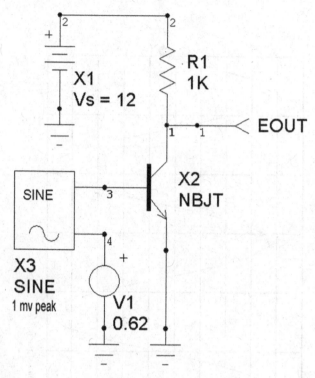

FIGURE 1.12
A BJT common-emitter gain stage schematic with 1 mA bias.

1.5 The Inverting Mode

Figure 1.14 shows the inverting closed-loop gain model of an op-amp-based design. The closed-loop gain at direct current (DC) (zero Hz) is simply the ratio of R2 to R1 with a negative polarity. Some advantages are (1) the input stage does not need to have large voltage compliance since the "virtual ground" sits close to 0 volts for any signal, (2) any non-linearity in the op-amp will be reduced by the ratio of G open loop to G closed-loop, (3) the inversion of waveform polarity may be needed for a stage in a feedback control system, and (4) several signals can be summed by simply feeding their signal currents into a series resistor, and then into the op-amp's virtual ground (node 5).

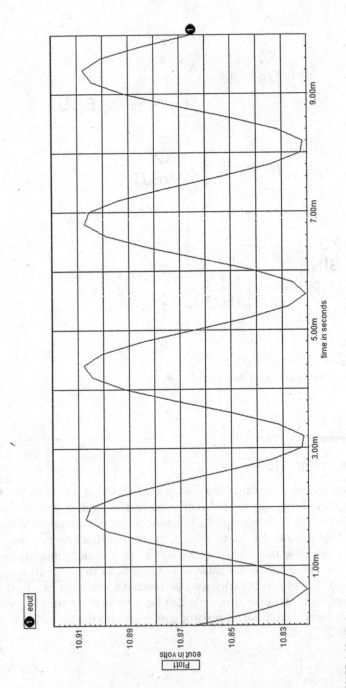

FIGURE 1.13
Output waveform of common emitter stage with 1 mA bias.

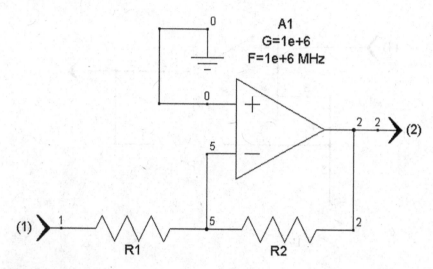

FIGURE 1.14
Inverting closed-loop gain model of op-amp-based design.

1.6 The Non-Inverting Mode and Its SNR Advantage over the Inverting Mode

Figure 1.15 shows the non-inverting application of an op-amp. The chief advantage of the non-inverting mode is its application to buffer circuitry, also called a voltage follower. It can present extremely high impedance to a source, minuscule impedance for driving a load, and gain extremely close to unity when R2 = 0 and R1 = infinity. Another advantage is signal-to-noise ratio. See Figure 1.16 for a view of noise gain of the inverter. Both circuits offer unity gain (but 180 degrees apart) but the voltage follower wins favorable SNR by 6 dB (see Figure 1.17).

1.7 The Operational Transconductance Amplifier

The operational transconductance amplifier is pictured in Figure 1.18, and is similar to the op-amp except that its output is a current rather than a voltage.

A BJT is a crude OTA, as shown in Figure 1.19. A once-popular OTA part was the RCA CA3080, and its schematic diagram is shown in Figure 1.20 and

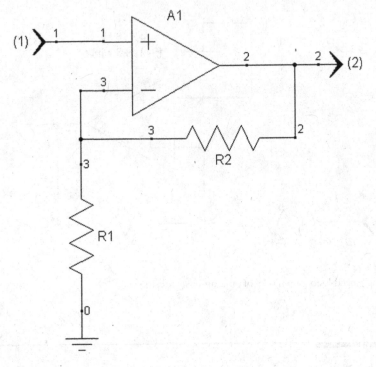

FIGURE 1.15
Non-inverting application of an op-amp (generalized gain).

corresponding graph of transconductance gain vs bias current appears in Figure 1.21. By switching the bias current off, the OTA output goes to a high impedance zero current state. Thus, its output can be combined with other OTA's outputs to form a multiplexer. Or the bias current can be continuously variable to form a voltage-controlled amplifier (VCA), assuming there are resistors at both the output and the bias control terminals. This application was perhaps its most useful and popular.

1.8 The Transistor as a Transconductance Amplifier

The circuit of Figure 1.22 is probably useless except to demonstrate the transconductance of a BJT versus V_{be} bias voltage. Ignore the circuitry to the right of Q1; it is simply a highpass filter to suppress the control signal superimposed on our desired waveform at Q1's collector. The emitter forward bias voltage was varied from 0.5 to 0.7 volts, and one can see the exponential

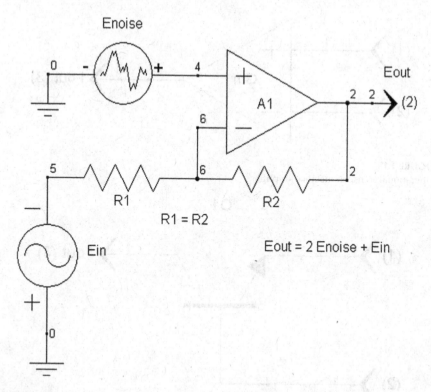

FIGURE 1.16
Inverting case unity gain with noise.

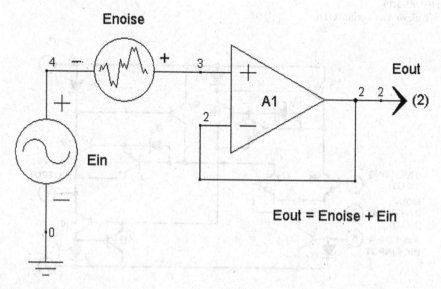

FIGURE 1.17
SNR advantage of unity gain non-inverting circuit and noise.

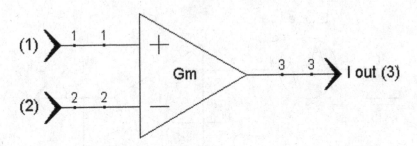

FIGURE 1.18
Operational transconductance amplifier.

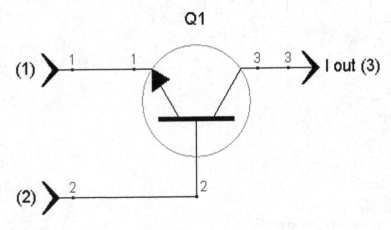

FIGURE 1.19
BJT viewed as a crude OTA.

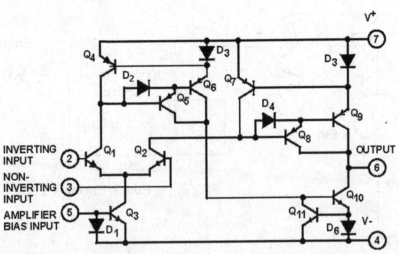

FIGURE 1.20
Schematic of a once-popular OTA (RCA CA3080).

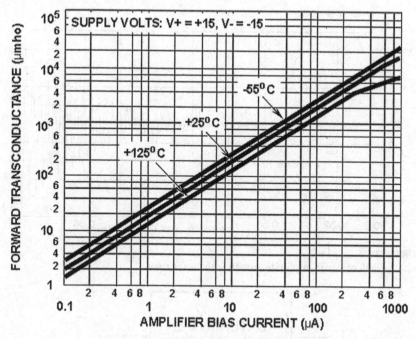

FIGURE 1.21
Graph of transconductance gain versus bias current (RCA CA3080).

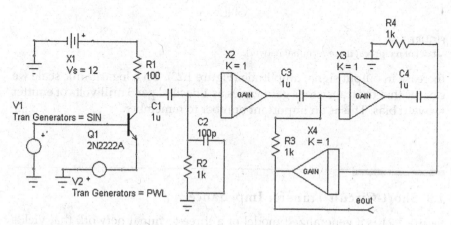

FIGURE 1.22
Demo circuit to show BJT transconductance gain versus bias.

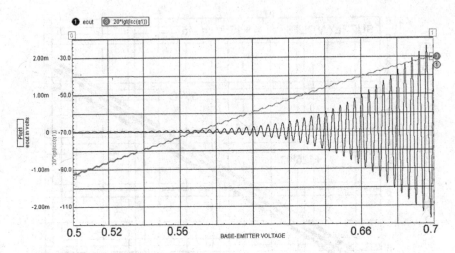

FIGURE 1.23
Resulting graph of signal amplitude growth with bias current.

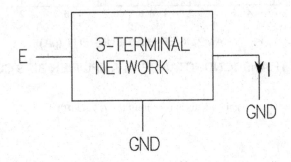

FIGURE 1.24
Model of reciprocal three-terminal network.

increase in output signal amplitude (Figure 1.23). On a logarithmic scale we can see that the transistor's gain varies as 1 deciBel per 3 millivolts of emitter forward bias. This is an important number to remember.

1.9 Short-Circuit Transfer Impedance

Figure 1.24 is a generalized model of a three-terminal network that yields a feedback current to an op-amp's inverting input from its output voltage. Since the op-amp's inverting input is essentially a virtual ground [3], the short-circuit transfer impedance sets the gain and response when the signal is applied to the virtual ground via a signal current.

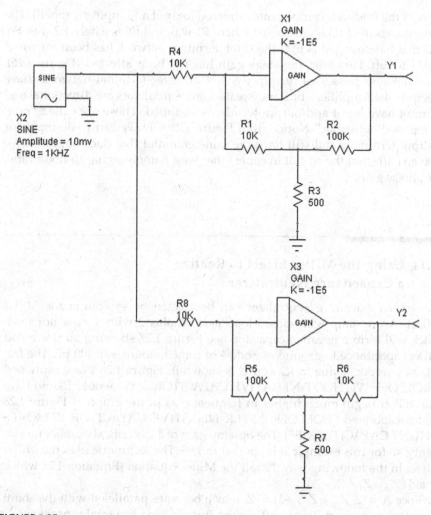

FIGURE 1.25
Upper half of schematic is an op-amp with three-terminal feedback, lower half is same network mirrored left to right.

1.10 Reciprocity of the Three-Terminal Feedback Network

In the upper half of Figure 1.25, we see that the closed-loop gain is set by a three-terminal feedback network involving R1, R2, and R3 and corresponding short-circuit transfer impedance, whereas the lower half of Figure 1.25

shows the feedback components mirrored (output and input swapped). The new component R5 is connected where R1 was and R6 is where R2 was. So, all that has changed is that the three-terminal network has been mirrored right to left. However, the stage gain has not been affected (Figure 1.26). The network possesses reciprocity. Not all three-terminal networks have reciprocity. Amplifiers, buffers, isolators, and circulators are directional and cannot have input and output terminals swapped. Those are called "non-reciprocal networks." Notice from Figure 1.26 that we can trade input for output terminals and still have the same gain. But that does not mean we haven't affected the circuit in some other way, namely signal-to-noise ratio, i.e., noise gain.

1.11 Using the Miller Effect to Realize a Capacitance Neutralizer

A clever capacitance neutralizer can be created by exploiting the Miller effect. An op-amp with closed-loop gain of plus 2 with a capacitor feedback will form a negative capacitor. See Figure 1.27, showing an unwanted stray capacitance to ground at node 4 of some amount, say 100 pF. The frequency response due to R2 and C is shown in Figure 1.28 (trace captioned "ROLLOFF WITHOUT NEGATIVE CAPACITOR"). We would like to have the rolloff begin much higher in frequency, as in the graph of Figure 1.28 (trace captioned "ROLLOFF AFTER NEGATIVE CAPACITOR REMOVES SHUNT CAPACITANCE"). The op-amp gain of 2 is a critical number for stability, so for this example it is spoiled to 1.95. The technique uses the Miller effect in the following way. Recall the Miller equation (Equation 1.2), which yields $Z_{in} = Z_f / (1 - A)$.

Since $A \approx 2$, $Z_{in} = Z_f / (-1) = -Z_f$ which becomes paralleled with the shunt capacitor to ground. Two parallel capacitors add to yield total capacitance.

This technique was actually applied to a HP product (new company logo = Keysight), the HP339 audio distortion test set, where the capacitance being neutralized was a stray capacitance to ground due to layout and other contributions (it was much less than 100 pF).

1.12 Viewing the Transistor as a Current Conveyor

In Figure 1.19, we viewed the BJT as a transconductance amplifier. The same model can be a current conveyor, wherein the emitter is fed by a current

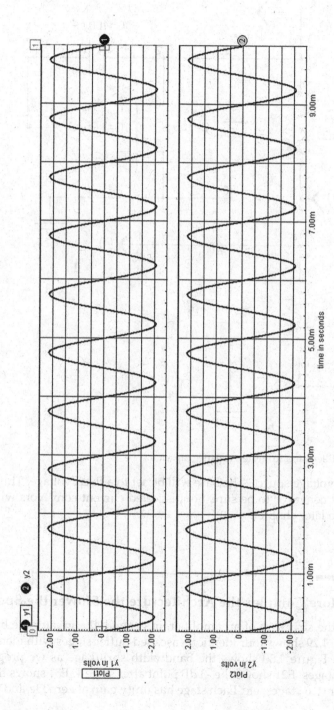

FIGURE 1.26
Outputs of both examples show the same amplitude.

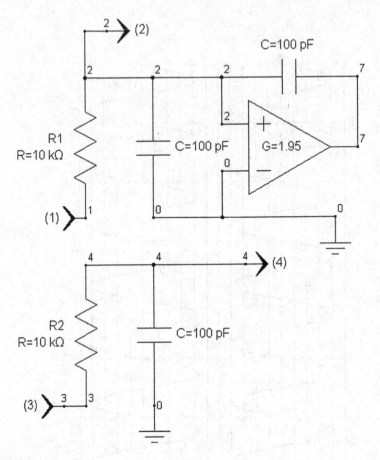

FIGURE 1.27
Exploiting the Miller effect to create a negative capacitor.

source, not a voltage source. The base will be set to a fixed voltage. This is a crude current conveyor to be sure. Sophisticated current conveyors will be discussed in a later chapter.

1.13 The More Complex the Architecture the Slower the Speed

The simpler the structure of an amplifier, in general, the greater the bandwidth. Figure 1.29 shows four identical cascaded buffer stages with identical 3 dB rolloffs. Figure 1.30 shows the bandwidth shrinkage as we progress through the stages. E21 shows the -3 dB point after stage 1. E31 shows the -3 dB point after two stages, etc. Each stage has unity gain at zero Hz and each

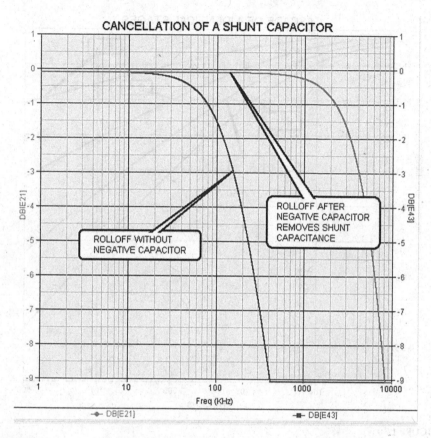

FIGURE 1.28

Frequency response plots before and after capacitance neutralization.

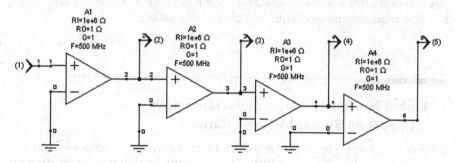

FIGURE 1.29

Schematic of 4 cascaded identical 3db bandwidth stages.

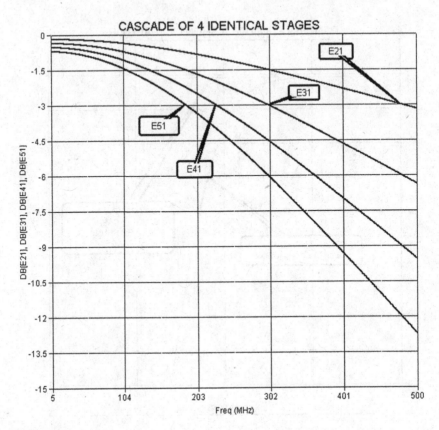

FIGURE 1.30
Plot of frequency response degradation through each subsequent identical stage.

stage has a 500-MHz bandwidth at the -3 dB point. Notice the compounding degradation after several stages. Not shown is phase shift which also deteriorates, requiring more dominant pole compensation.

1.14 Shot Noise and Transconductance and Impact on Signal-to-Noise Ratio

When a transistor is operating at a collector current of "I_{bias}" amperes, and we are observing a bandwidth of "BW" Hz, there will be shot noise according to the following equation:

$$I_{shot} = \sqrt{2qI_{bias}BW} \qquad (1.5)$$

A computer program written in BASIC can streamline the calculations (translation to other code is left to the reader):

```
10 REM  THIS PROGRAM CALCULATES SHOT NOISE ASSOCIATED
   WITH A BIAS
20 REM  CURRENT IDC, AND RETURNS TO THE USER ISHOT, RMS,
   FOR A
30 REM  USER-DEFINED BANDWIDTH IN HERTZ.
40 REM
50 CLS
60 PRINT:PRINT:PRINT:PRINT
70 PRINT "***********************************************"
80 PRINT:PRINT
90 PRINT "       SHOT NOISE CALCULATION PROGRAM
   'SHNOISE'"
100 PRINT
110 INPUT "DC BIAS CURRENT, IDC (AMPS) =";ADC
120 INPUT "BANDWIDTH IN HZ, BW     =";BW
130 PRINT
140 ASH=SQR(2*1.59E-19*ADC*BW)
150 PRINT "SHOT NOISE RMS IN BW =";ASH
155 PRINT "SHOT NOISE IS ";20*LOG(ASH/ADC)/LOG(10);"
   DECIBELS BELOW IDC"
156 PRINT
157 INPUT "TRY AGAIN (Y/N) ";A$
158 IF A$="Y" THEN GOTO 100
159 IF A$="y" THEN GOTO 100
160 STOP:END
```

To verify that the code works properly try I = 10 mA, BW = 1 KHz; then the resultant shot noise should be 1.78 nanoamps RMS. To achieve a SNR of 80 dB in that bandwidth, therefore, we must have a signal current of 10,000 times the shot noise value, or 17.8 microamps RMS.

Since G_m of a BJT increases directly as bias current, but shot noise increases as the square root of said current, the best SNR is obtained at high bias currents. In addition, the gain-bandwidth product increases at larger G_m values. Thus, op-amps with high GBW are generally operated at higher front-end bias current.

1.15 Resistor Noise

A resistor noise model looks like an ideal resistance in series with a noise voltage (generated by the resistor). This noise voltage is usually expressed in volts per root-Hz. A good number is the benchmark of 4 nanovolts per root-Hz for a 1 K-ohm resistor at room temperature. The noise voltage varies as the square root of the resistance, so to get 1 nanovolts per root-Hz, the resistor must be about 60 ohms. Some op-amps have less than 1 nanovolt per root-Hz input referred noise, so we must choose suitably low resistance values at the op-amp inputs in order to hold circuit noise to a minimum.

The equation for resistor noise voltage density is:

$$E_n = \sqrt{4kTR} \qquad (1.6)$$

where

k = Boltzmann constant = 1.38×10^{-23}

and T = absolute temperature in degrees Kelvin (room temperature approximately equals 300 degrees) and R = resistance value in ohms, and the total noise voltage is proportional to the square root of bandwidth and expressed as volts per root-Hz.

1.16 The Darlington Configuration for RF Amplification

Refer to Figure 1.31, showing the Darlington configuration of active devices (in this example they are BJTs) used in many gain stages, at high frequencies. Because of its simplicity, bandwidth is high and many products using this technique exist, especially in silicon-germanium (SiGe) BJTs with f_t above 50 GHz. Thus, for many applications, the parasitic reactances of the active devices can be considered negligible at operating frequencies up to 1 GHz.

Figure 1.32 (upper plot) shows the output waveform into a 50-Ohm load when the circuit of Figure 1.31 is driven by a 50-Ohm source at 100 mV peak. The output is 650 mV peak at 50 MHz, which agrees with the small-signal gain of 6.6 at 50 MHz (lower plot). Note that the gain asymptotically approaches 7.2 (equals 17.1 dB) above 1 GHz.

Figure 1.33 shows the simplification of the circuit if we make Q1-Q2 appear as a voltage-controlled current source whose transconductance gain would have been set by Q2's collector bias current but is reduced to the value of (1/R4) (in this example = 178 mmhos). Ports 1 and 2 are matched to 50 Ohms.

Figure 1.34 shows the input match to 50 Ohms (s11) and output match (s22) as well as the gain (s21).

S11 is usually given as deciBels of reflection (similar to return loss). The gain and input impedance are established by R1 and R5, and somewhat by R4 (in Figure 1.31) and the bias current in Q2's collector. We have assumed L1 is so large as to present a sufficiently high impedance to be ignored, and all coupling capacitors' impedances to be low enough to be regarded as short circuits to AC.

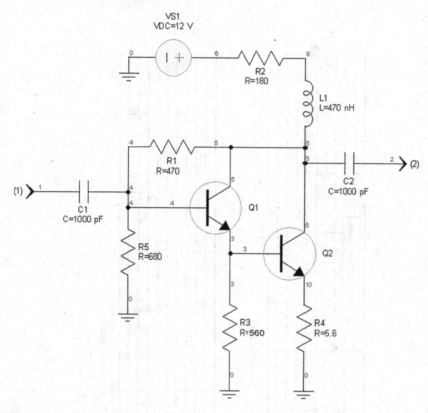

FIGURE 1.31
BJT Darlington RF amplifier example.

In an integrated circuit transistor Q2 is physically larger than Q1 out of the need to operate at a higher current. Being essentially class A, this Darlington gain block cannot deliver more load current than its bias current. A signal of greater amplitude will be severely compressed.

1.17 Non-Small-Signal Amplifiers

1.17.1 Class C

The advantage of class C amplification is efficiency compared to lower classes A and B. However, class F is even more efficient and we will show examples of both. Class C operates the transistor in conduction for only less than 50% of the waveform period. (In Class B the transistor is active for exactly 50% of the period.) This means that the output waveform must be loaded with a tuned circuit. Furthermore, for each fraction of a period that the device is

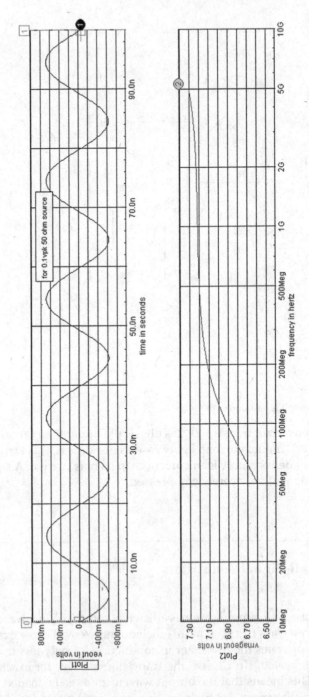

FIGURE 1.32

Output waveforms of Darlington RF amplifier (see text).

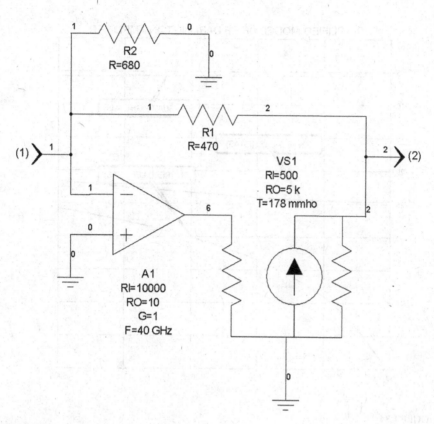

FIGURE 1.33
Simplified "ideal" model of RF Darlington amplifier.

active, it must deliver a large "poke" of current to the tuned circuit. Because the device is active for so much less than a half period, efficiency is high. See Figure 1.35 and notice that Q1's collector is fed to a tank circuit with a 1500-Ohm load (R3). It is not very useful to have such a high impedance externally, but 1500 Ohms establishes a large signal at the tank and sets the bandwidth of the tank narrow for best signal filtering. However, the 1500-Ohm load can be power-matched to a 50-Ohm load as shown in Figure 1.37, changing a single capacitor C3 (we will discuss this matching technique in a later chapter). Figure 1.36 shows the output waveforms of this class-C example operating into a final load of 1500 Ohms and the value of C3 is large. Notice that the sine wave's negative peak extends to nearly 0 volts, so the transistor is operating at full saturation on negative output signal peaks. Figure 1.38 shows the waveform after matching to 50 Ohms. Notice that the signal amplitude has reduced but because of the 50-Ohm loading, the signal power has remained at 47 milliwatts. Thus, the matching technique imitates a transformer.

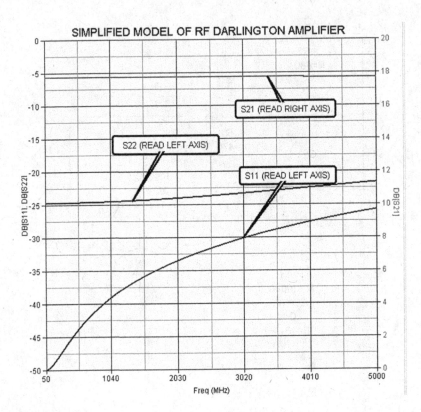

FIGURE 1.34
Plots of gain and input-output reflection of RF Darlington.

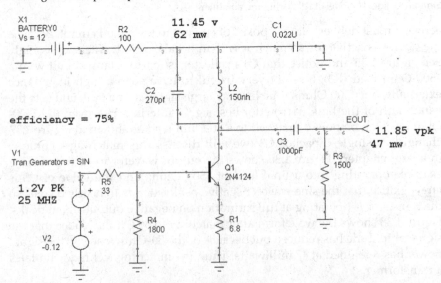

FIGURE 1.35
Class C example schematic with 1500–Ohm load.

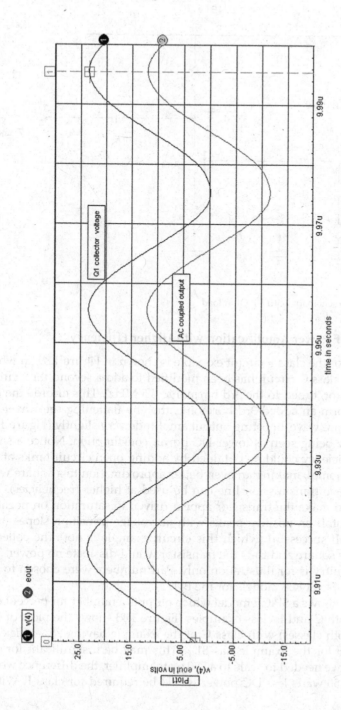

FIGURE 1.36
Output waveform at Q1 collector and with DC blocking.

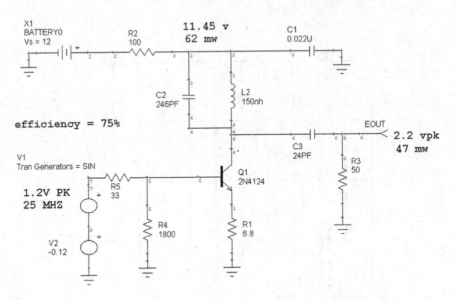

FIGURE 1.37
Class C example schematic with 50–Ohm load.

1.17.2 Class F Power Amplification with Higher Efficiency

A schematic of the class F circuit example is shown in Figure 1.39, in which the previous class C circuit has been modified to add a second tank circuit in Q1's collector, tuned to the 3rd harmonic (75 MHz). This affords the collector waveform an added 3rd harmonic, thereby flattening the waveform toward a square wave, boosting output amplitude very slightly (Figure 1.40) but primarily being seen as lower DC power consumption. Notice a small benefit in efficiency could be obtained by adding open circuit tanks at 5th and 7th harmonics, making an even better approximation to a square wave (a quarter-wave transmission line can be used at higher frequencies). The objective is to make the transistor appear driven to saturation on negative peaks and cutoff on positive peaks with no voltage transition slopes (note Figure 1.41). If successful (which this circuit example is not), the collector waveshape is square and thus the transistor would dissipate no power. The example circuitry is for illustration only. Part numbers were chosen to use readily available SPICE models for the BJT.

Figure 1.42 shows a SPICE model which outputs a number for percent efficiency for both C and F class examples (Figure 1.43 shows the plots of efficiency for both classes) with class C of the example having 75% efficiency while class F for the example has 81%. This may be insignificant for this circuit but if we needed to scale to a 770 watt amplifier, the difference would be that over 50 watts less DC power would be required for class F. With a

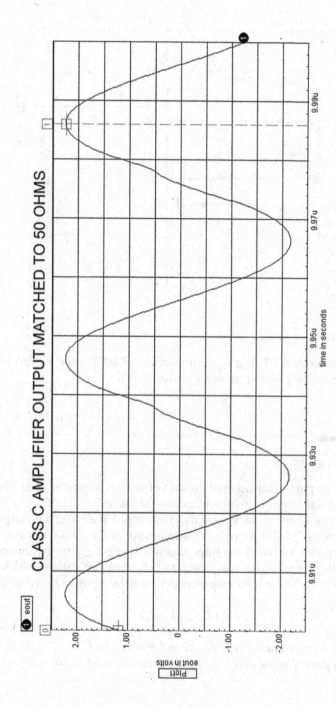

FIGURE 1.38
Output waveform at 50-Ohm load match.

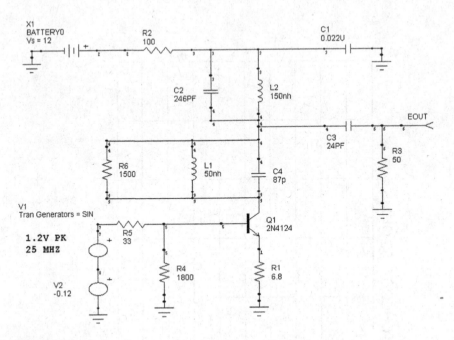

FIGURE 1.39
A schematic of the class F circuit.

field-effect transistor (FET) (e.g., gallium nitride – GaN)-based design better performance may be possible at lower power.

1.18 Gyrators

An example of a particular non-reciprocal network is a gyrator [7, 8]. Figure 1.44 shows a crude example which uses two transconductance amplifiers of opposite polarities back to back. One (G_{m2}) has input at Z1 and has output at Y2 and the other (G_{m1}) has input at Y2 and output at Z1. This constitutes an impedance inverter, i.e., at Y2 the impedance will appear to be the inverse of Z1 scaled by the reciprocal of the product of the two Gm values (which can be equal if desired). The impedance seen at Y2 will be equal to Z_2. What is Z_2?

$$Z_2 = 1/(G_{M1}G_{M2}Z_1) \tag{1.7}$$

Gm values are in mhos and Z values are in Ohms.

Let Z1 be a purely capacitive element (but not limited to a capacitor). We can state that

$$Z_1 = 1/(j\omega C) \tag{1.8}$$

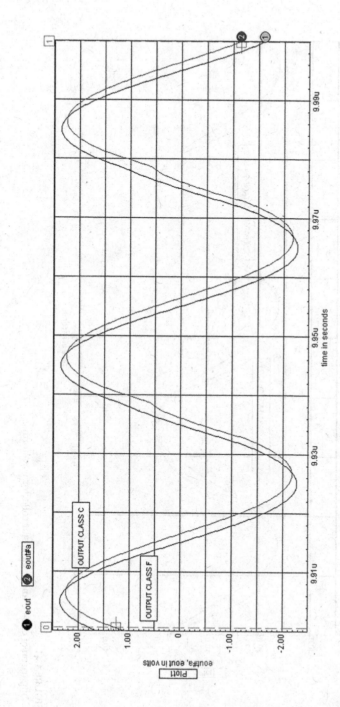

FIGURE 1.40

Comparison of class C and F output waveforms.

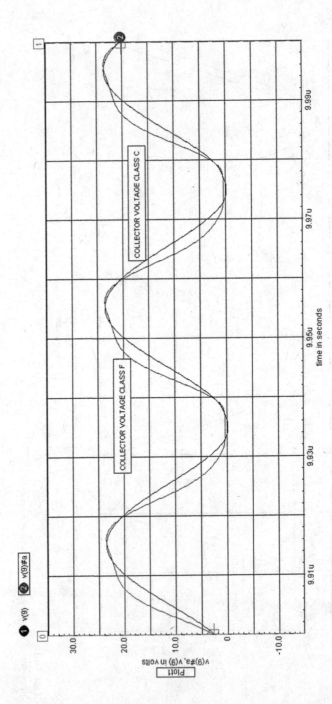

FIGURE 1.41
Comparison of class C and F transistor collector waveforms.

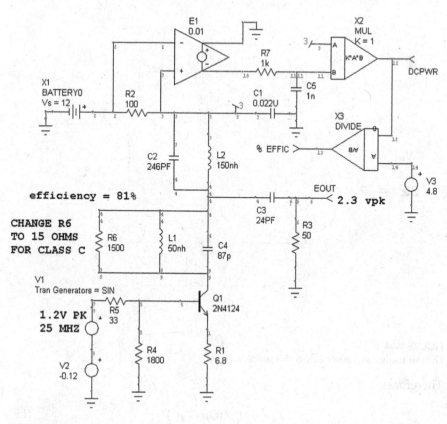

FIGURE 1.42
A more complete efficiency-calculating model schematic.

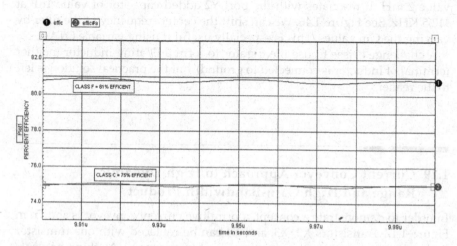

FIGURE 1.43
Plots of class C and class F efficiency numbers in percent.

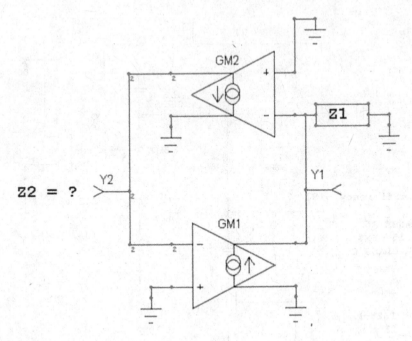

FIGURE 1.44
Gyrator using two transconductance amplifiers.

Therefore

$$Z_2 = j\omega C / (G_{M1}G_{M2})$$ (1.9)

For the example of Figure 1.45, we have created an inductor at port Y2 of value 2 mH. It resonates with the port Y2 added capacitor of value 1nF at 112.5 KHz. See Figure 1.46. We can shift the center frequency of the filter by varying the Gm values. This is especially useful if using variable OTAs.

A challenge arises to use the gyrator to form a floating inductor (neither terminal of inductor is connected to ground), but the practical solution is left to the reader.

1.19 Current Conveyor Approach to High Dynamic Range and High Gain-Bandwidth Product

In order to demonstrate a concept, a primitive current conveyor is shown in Figure 1.47. Transistors X2, X3, and X4 can be replaced with one transistor with an emitter area three times the size of transistor X1. Nothing prevents

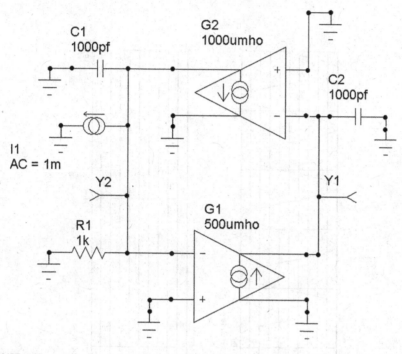

FIGURE 1.45
Creating an inductor from a capacitor using an impedance inverter (gyrator) forms a 112.5-KHz bandpass filter of unity gain.

using transistors other than BJTs. V_{be} matching is required of all four transistors, something better achieved in an integrated circuit. *I1* is an input current of 5 mA peak at 100 MHz and I2 is 10 mA to assure that the transistors remain active in this unipolar circuit. C1 was needed in SPICE simulation for stability. See Figure 1.48 for performance predicted by SPICE.

Current conveyors have an important quality: the absence of Miller effect when loaded in a virtual ground.

When signal voltage swings are minimal, such that the circuit has no voltage gain, it can have no Miller effect.

A unity gain current conveyor with superior input virtual ground (close to zero impedance) is shown schematically in Figure 1.49. It also has extremely high rejection of the input baseline voltage Y, assuming the input current source at X has infinite parasitic shunt impedance (high compliance). Its performance is shown in Figure 1.50. Current conveyors can be cascaded for high current gain before their outputs need conversion to a voltage signal. The lower plot in Figure 1.50 is the baseline waveform Y, which is nearly totally ignored by the circuit as desired (desired waveform is upper plot). Current gain as shown is unity, but can be increased by equally increasing the scale of the emitter areas of the two output transistors.

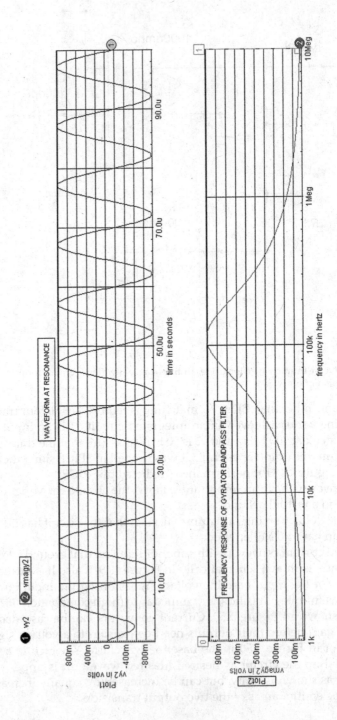

FIGURE 1.46

Gyrator bandpass filter example; transient and frequency response.

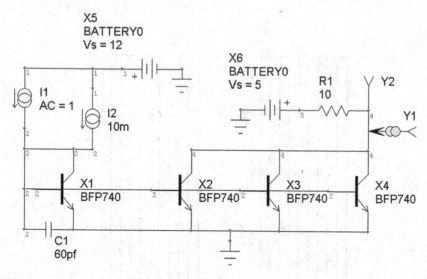

FIGURE 1.47
A primitive current amplifier/conveyor.

1.20 Linearity

SPICE provides a way to measure non-linearity of a circuit such as the current conveyor. A DC sweep can be performed, then the resultant ramp at the circuit output can be differentiated. If linearity is perfect, the result should be a constant slope value, i.e., a straight horizontal line. Any kinks or bumps represent non-linearity. Refer to Figure 1.47 and apply a DC sweep to *I1*, then observe the output result (Figure 1.51 upper plots). Since the stimulus is perfect (generated by SPICE) the resulting slope should be a constant horizontal line (Figure 1.51). Now look at the lower plots showing the output current and its derivative which varies by 1.3%. A non-linearity of 1.3% is not unexpected, since the circuit is mostly open loop (no feedback). The poor linearity of most current conveyors (albeit better than this example) may rule them out for many applications, which favor voltage gain and voltage amplifiers which mostly use feedback or feed-forward correction and typically achieve distortion levels of -90 dB or better. For example, see Figure 1.52 which shows the SPICE model of a voltage follower using the uA741 op-amp. The source (V1) DC sweep of Figure 1.53 shows a unity gain slope and a slope departure (derivative in SPICE) of 0.00063% (–104 deciBels). Of course, these comparisons are made at zero Hz. At high frequencies, the effectiveness of negative feedback shrinks as the gain-bandwidth limitation causes less available gain

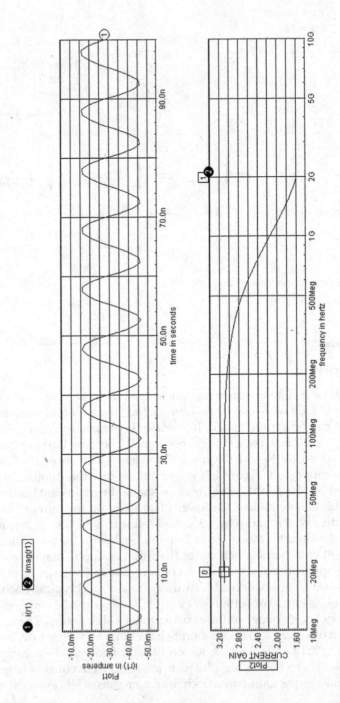

FIGURE 1.48
Primitive current conveyor SPICE simulation results.

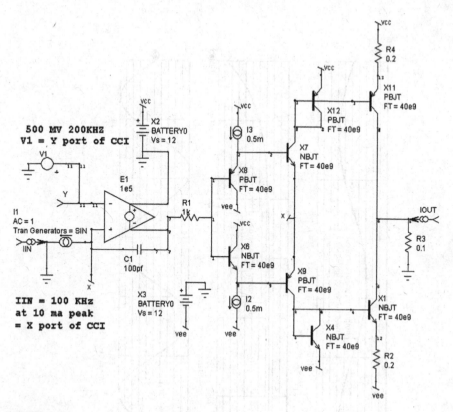

FIGURE 1.49
Improved current conveyor gain (CCI) incorporates an op-amp.

to be fed back, which yields less corrective behavior, until perhaps the linearity issue improves for the current conveyor approach.

1.21 Physical Layout and Parasitics Caused by Layout

Component placement and printed circuit board (PCB) layout is critical to avoid the Miller effect even in an IC. For example, a cascode of two transistors is meant to avoid the Miller effect of the bottom device, but this can be degraded if layout causes voltage coupling from the top device's output to the bottom device's input. See Figure 1.54, which shows just such capacitive coupling due to sloppy layout. Figure 1.55 shows the reduction in bandwidth caused by the poor layout around the cascode. FET M1 by itself has a frequency response (when driven from 1K-Ohm source and loaded at drain by 1K Ohms) as shown in the middle trace. When transistor M2 is added in

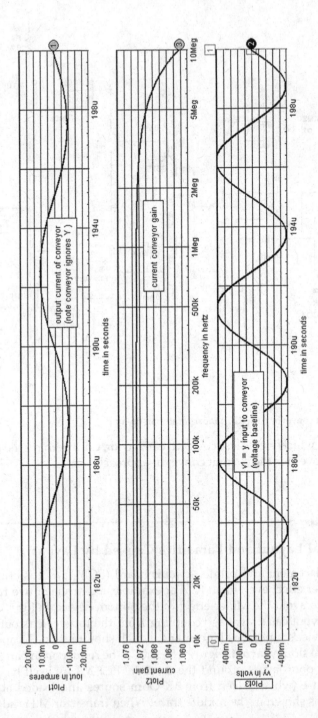

FIGURE 1.50
Graphic results of improvedCC1 performance.

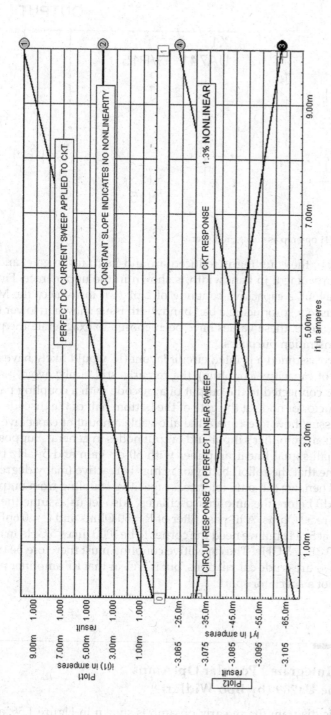

FIGURE 1.51
Stimulus (top) and CKTresponse to DC sweep.

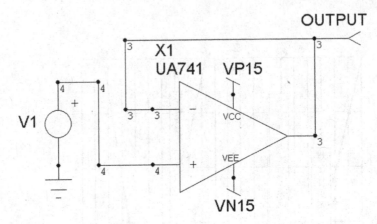

FIGURE 1.52
SPICE model of 741 op-amp as voltage follower.

cascode fashion, the coupling is reduced around its drain to gate, and the beneficial improvement in bandwidth is shown in the upper trace. Finally, due to poor layout, a capacitive coupling of 2 pF (C1) introduces the Miller effect to the cascode, producing a bad bandwidth issue shown in lower thinner trace. Of course, these values have been chosen to exaggerate the problem for demonstration purposes.

Although we assumed a printed circuit layout (IC would likely have miniscule values of capacitive coupling), the remarks still apply: avoid placing any conductor connected to the output of a cascode with a coupling path to another conductor or node at the gate of the bottom half of the cascode.

One more issue: how to estimate the allowable amount of capacitive coupling between an amplifier's input and output nodes in general. Suppose we have a RF small-signal linear amplifier with 60 dB gain and 5 GHz bandwidth. Assume that the rolloff behavior is highly reactive (many degrees of phase shift). Then, we can only tolerate a capacitive coupling from output to input of −60 dB before the amplifier oscillates. Also let us assume the port impedances are 50 Ohms. A hi-pass filter of R = 50 Ohms and C = 0.6pF will have a corner at 5GHz but we need the corner to be 1000 times 5 GHz in order to be down 60 dB at 5 GHz. The capacitive coupling must therefore be under 0.0006 pF. These are crude calculations, but it is clear that RF amplifier physical layout is not a spectator sport.

1.22 Early Integrated Popular Op-Amps and the Ua709 (by Bob Widlar)

The schematic diagram for an early op-amp is shown in Figure 1.56, and it is the uA709 designed by analog pioneer Bob Widlar. All capacitance values

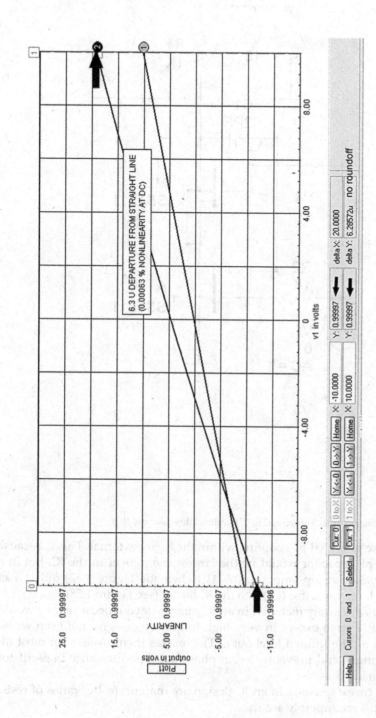

FIGURE 1.53
Linearity of 741 op-amp as voltage follower.

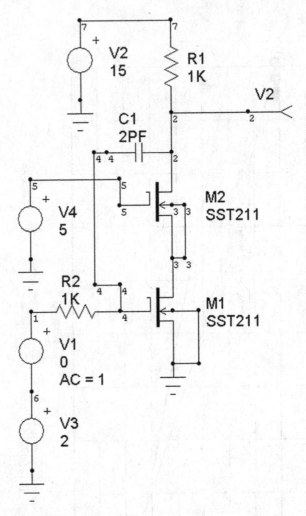

FIGURE 1.54
Poor layout caused capacitive coupling (C1) around this cascode.

above those provided by coupling within the IC are external. This is because even a 30-pF capacitor would be the largest component in the IC. But in a later generation of op-amp, the uA741 (schematic Figure 1.57) did in fact bring this large capacitor (C1) into the IC layout (see Figure 1.58).

Inductors are rarely included in an IC internal layout because they would be physically large except in very high-frequency designs, but even worse their Q would be limited to about 8. This makes them useless for most filter applications and prevents decent phase noise performance in oscillator applications.

In most cases, resistors in an IC design are inaccurate, but ratios of resistance values are superbly accurate.

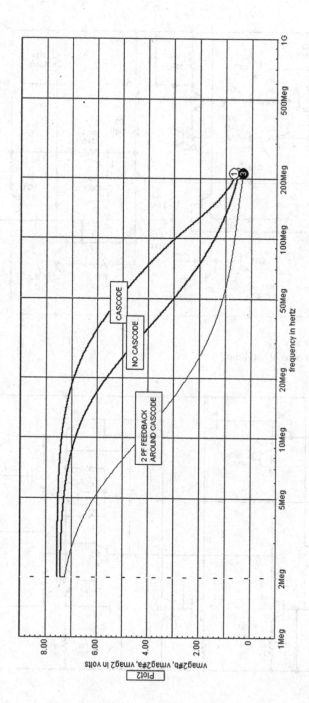

FIGURE 1.55

Frequency response before cascode, after cascode, and after sloppy layout.

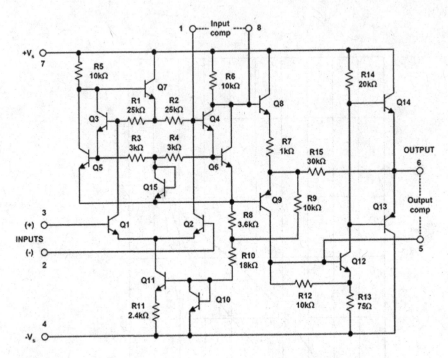

FIGURE 1.56
Schematic diagram of ua709 op-amp.

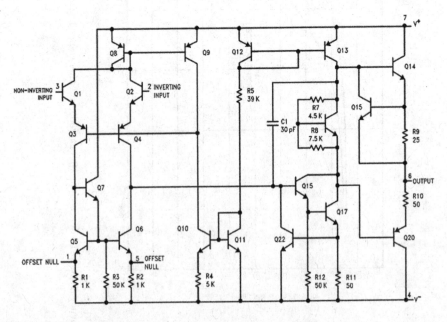

FIGURE 1.57
Schematic diagram of later design ua741 op-amp.

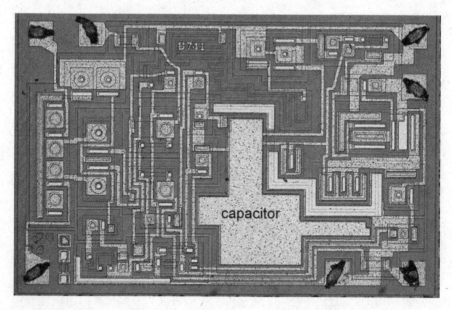

FIGURE 1.58
Layout of ua741 op-amp chip.

Many times, the transistors are tiny by contrast to other components, encouraging the engineer to use them liberally. Matching of parameters between transistors is also superb.

1.23 Transistor Issues

As the need for faster circuitry continues, transistor dimensions have grown smaller, and f_t's have gotten larger. Recall that f_t is the frequency at which a BJT transistor's beta drops to unity. Semiconductor material has also undergone improvements, for example, SiGe, gallium arsenide (GaAs), and GaN. Older BJT processes have spawned f_ts of around 3 to 6 GHz, whereas newer SiGe processes have achieved f_t greater than 60 GHz. As stray capacitances have gotten smaller with the advancing technologies, bandwidths have improved. Unfortunately, in some cases, breakdown voltages have gotten lower. Thus, many high-speed op-amps are usable only at lower supply voltages. Also, many fast devices have 1/f noise corner frequencies that are large, e.g., 1 to 10 MHz is not unusual.

2

Transimpedance Amplifiers for Low Noise

2.1 Introduction

For signal sources with high impedance, the best signal-to-noise (SNR) may be obtained from a transimpedance amplifier. Feedback resistor bandwidth and input node shunt capacitance are critical determinants of performance.

Transimpedance amps are thus used to amplify the signals from photodiodes and avalanche photo diodes (APDs).

2.2 Motivation

When the source has a high impedance, it is often advantageous to apply a different type of amplifier that converts the source current into an output voltage as noiselessly as possible. Such an amplifier is the transimpedance amplifier [4, 5, 6].

The transimpedance amplifier model is shown in Figure 2.1 with its parasitic input capacitance (from many sources, such as transistor input capacitance, transducer capacitance, photodiode, layout, etc.), its feedback resistance and parasitic feedback capacitance, and its finite inverting gain A0. The input is from a high-impedance current source, I_{IN}, such as a photo diode.

2.3 Resistor Bandwidth

Amplifier gain is A_0; output is in volts. The amplifier is assumed to be noiseless and drawing no input current, and having an infinite bandwidth for the

DOI: 10.1201/9781003088547-2

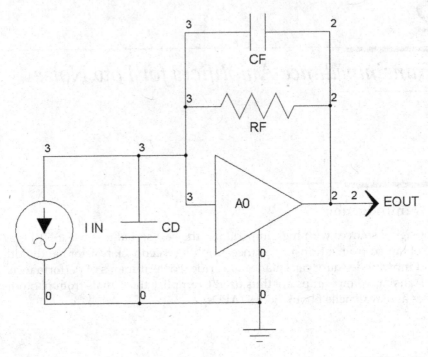

FIGURE 2.1
Transimpedance amplifier model.

purpose of illustrating some important behaviors. The transimpedance gain
is thus:

$$A_z = \frac{E_{out}}{I_{in}} = \left[\frac{sC_d}{A_0} + \frac{1+sR_fC_f}{R_f}\left(\frac{1}{A_0}-1\right)\right]^{-1} \tag{2.1}$$

If we force all capacitors to have a zero value, the above equation simply
becomes:

$$A_z = \left[\frac{1}{R_f}\left(\frac{1}{A_0}-1\right)\right]^{-1} \tag{2.2}$$

Thus, for A_0 very large and negative, $A_z = -R_f$.

Properly designed, the principle noise contributor in this model is the
feedback resistor R_f. If too small in value we would have too little output
signal and if too large, we have too little bandwidth (since the resistor has
parasitic parallel capacitance). In most applications, like fiberoptic communication systems, the receiver front end is a transimpedance amplifier specially
designed for the bandwidth and SNR required. We would like as large a
resistor value as possible, because its noise will be bigger than the amplifier's input noise, and its effect on signal is to make it as large in voltage as

possible. The input current of the amplifier is assumed to be negligible compared to signal current. It is also assumed that we need a bit error rate (BER) less than 10^{-9}. Then the feedback resistor must be larger than some number of Ohms, expressed as follows:

$$R_f > 2.304 \times 10^{-18} \left[\frac{BW}{\left(I_{Smin,pk} \right)^2} \right] \tag{2.3}$$

where

BW = bandwidth of final result

I_S min, pk = lowest peak input signal current to be processed

For BW = 50 MHz and minimum peak source current = 10 nA, the feedback resistor must be larger than 1.15 MegOhms. But to obtain 50 MHz bandwidth, the resistor parallel capacitance must be less than 0.000277 pF. Such a resistor does not exist. A thick film resistor has about 0.3 pF of shunt capacitance, a thin film resistor is considerably better, and a thin film on quartz or sapphire resistor can have as little as 0.02 pF.

So, for this example our source must not be a simple photodiode, but an APD. Such a detector can operate easily at a multiplication factor M = 20. In that case, the required minimum feedback resistor will be 2.8 KOhms, which must exhibit less than 1.13 pF, inclusive of layout. So we can now deliver a receiver front end with adequate bandwidth and sensitivity, provided that the amplifier noise does not "out-shout" the resistor noise. The noise of a 2.8 KOhm resistor is 6.7 nanovolts per root Hz. We can certainly find a transistor with such low voltage noise and current noise less than 1 nanoamp in 50 MHz. However, two other transistor parameters may bite us: 1/f noise and gate to source capacitance. A GasFET meets the low capacitance and low input current requirements but has a 1/f noise knee as high as 10 MHz. This puts a bow in the noise density curve that can be mistaken for a bandwidth issue, especially if no APD is used.

2.4 Cascode Input Stage

Since we desire very high transimpedance gain in as large a bandwidth as possible, the first stage of the amplifier must minimize the Miller effect. Such a first stage is the "cascode," not to be confused with "cascade." We have already described such a topology in Section 1.22, and Figure 1.54, but with the desired value of C1 close to zero. The signal input node must also be as low capacitance as possible, including the capacitance of the photodetector (or other current source) and layout for example. A value much less than 2 pF is desired. That is because after negative feedback which encloses the input node said node capacitance factors into the noise gain versus frequency.

2.5 Tricks when Bandwidth Is Insufficient

Assuming that the components have been optimally selected, one can artificially improve the bandwidth of the transimpedance amplifier by using a three-terminal network for the feedback, rather than a two-terminal one (the feedback resistor and its shunt capacitance). However, although this technique makes frequency response flatter, it ruins the noise gain. There is no clever simple way to flatten frequency response that does not impact SNR. One could try a capacitance neutralizer like that previously described in Section 1.12 but it would likely have unreasonable demands imposed on its bandwidth and noise performance.

2.6 Input Node Capacitance Issue Drives Noise

Refer to Figure 2.2. Capacitor C5 is a primary issue. (Another is feedback resistor shunt capacitance.) Figure 2.3 shows the gain versus frequency (lower plot) of the realized circuit at very nearly 100K, and its active region is shown at the upper plot. Figure 2.4 shows the resulting noise density curve and cumulative noise.

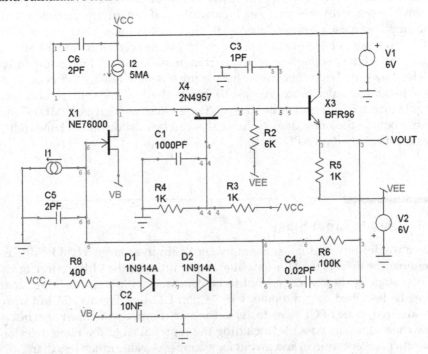

FIGURE 2.2
Schematic of possible 25 MHz transimpedance amplifier.

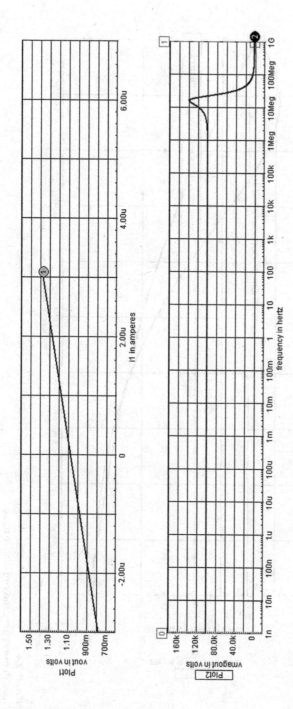

FIGURE 2.3

Plots of frequency response and linearity of transimpedance amplifier of Figure 1.60.

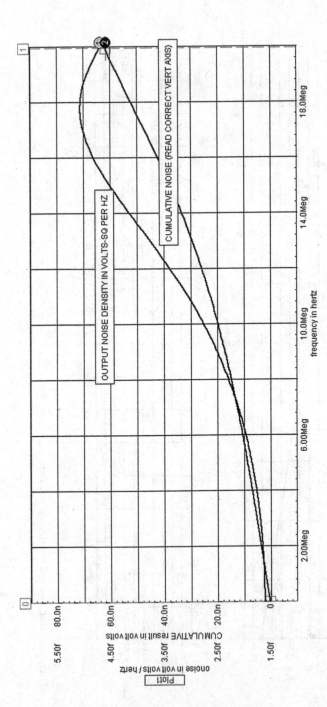

FIGURE 2.4

Noise density curve and cumulative noise.

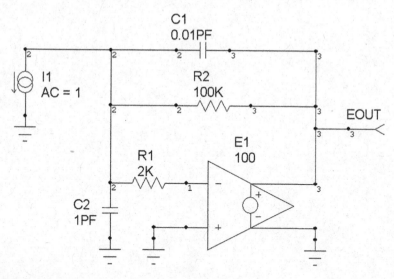

FIGURE 2.5
Idealized transimpedance amplifier.

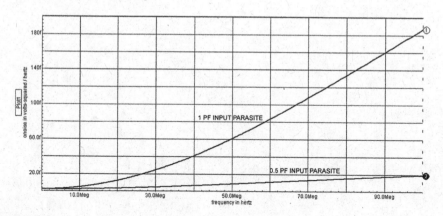

FIGURE 2.6
Comparison of noise curves with 0.5 pF and 1 pf for C2.

Figure 2.5 shows an idealized transimpedance amplifier. Capacitor C2 increases noise gain of the amplifier while R1 models a transistor input noise of 5.6 nanovolts per root Hertz. In reality, using the newest processes, this noise source level is much lower (reaching 1 nanovolt per root Hz). Figure 2.6 shows the output noise for this ideal circuit with 1 pF input capacitance and with 0.5 pF. The impact on noise is remarkable.

3

Voltage-Controlled Amplifiers

3.1 Introduction

In this chapter, the voltage-controlled amplifier (VCA) is described and its principles are developed. In all cases described, a log-antilog circuit approach is taken. The first example is the Blackmer VCA, successfully deployed in dbx trademarked commercial audio products, that have nearly a 120-deciBel dynamic range. But bandwidth is extremely limited to essentially the audio frequency range. When that disadvantage is intolerable, the Talbot VCA circuit which has much higher bandwidth but much lower gain control range is suggested as a solution and described.

3.2 Log/Antilog Approach

Almost all VCAs or attenuators operate on the log/antilog approach. Recall that the logarithm of two multiplied quantities is equal to the log of the first plus the log of the second. This behavior is exploited by analog multipliers by making the first quantity a direct current (DC) bias and the second quantity the signal. Then the resulting logarithm is the log of the product, which can be "antilogged" to get the actual mathematical product. The scaled waveform at the output of this process contains some error (distortion) depending on how accurate the log and antilog processes are, but typically using well-matched transistors, it is under 0.2%.

There are two types of multiplier: 2-quadrant and 4-quadrant. The ones we will discuss in this section are both 2-quadrant. A 2-quadrant multiplier is only capable of scaling the signal port amplitude which can be either polarity, according to the value of the control port but ignoring its polarity in determining the sign of the multiplier output. A 4-quadrant multiplier can accept the control port polarity and use that polarity in determining the sign of the multiplier output.

DOI: 10.1201/9781003088547-3

3.3 Blackmer VCA

In 1971 David Blackmer [13, 14] filed a USA patent application and on January 30 of 1974 was issued number 3,714,462 for a multiplier circuit which became ubiquitous in the audio industry. It has been called the Blackmer VCA, and Blackmer was issued a later patent for an improvement involving twice as many transistors (USA patent 4,403,199). We have taken the liberty to make a Simulation Program with Integrated Circuit Emphasis (SPICE) model of one possible implementation appearing in Figure 3.1 using transistors Q2 and Q3 to perform the log function (in conjunction with op-amp X1) and Q1 and Q2 to perform the antilog in conjunction with op-amp X2. The voltages V5 and V6 serve to provide crossover bias so the transistor quad can never be inactive, and they set the noise floor. They are shown as fixed voltages in this simulation although they are in actuality a transistor and some resistors in order to temperature track the other transistors. For simulation purposes, we assumed all transistors are matched, and are not necessarily the part numbers one would choose for the final circuit. In fact, the entire circuit has been implemented on a thick film hybrid to assure temperature tracking of the

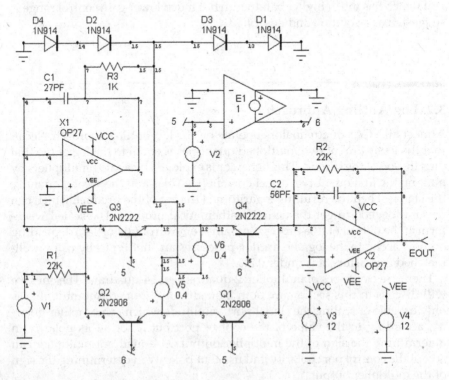

FIGURE 3.1
Blackmer VCA schematic.

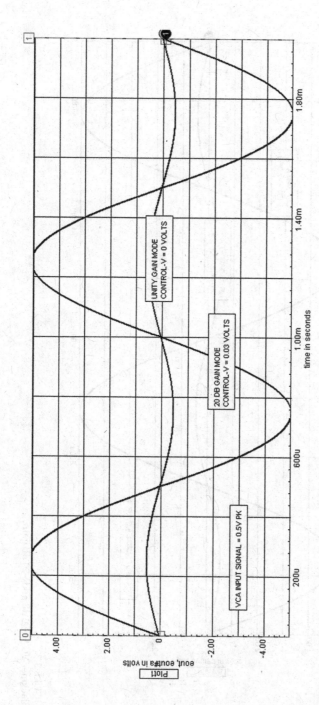

FIGURE 3.2
Blackmer VCA at 0 dB and +20 dB gain.

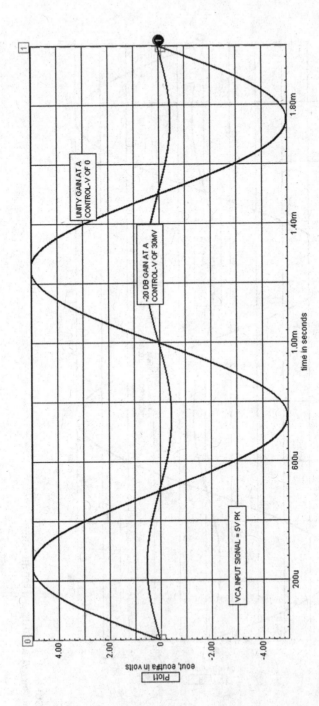

FIGURE 3.3
Blackmer VCA at 0 dB and −20 dB gain.

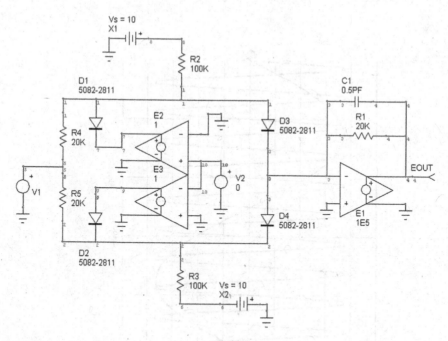

FIGURE 3.4
Talbot VCA schematic.

components. Ultimately, it has been redesigned for an integrated circuit (IC) marketed by THAT Corporation (yes, that's their name). The performance is shown in Figures 3.2 and 3.3. The drawback to this circuit is its 30 KHz or so bandwidth limit and the variation in accuracy at extremes of the control voltage. As shown its gain/loss is proportional to 1 dB per 3 millivolts of control voltage V2 and inverter E1. For V2 = 60mV, the gain is -20 dB and for V2 = −60 Mv, the gain is +20 dB and so on. The circuit has been used in audio equipment to cover a > 100 dB dynamic range.

If more bandwidth is needed, the Talbot VCA (next topic) can be applied to yield > 50 MHz bandwidth at the expense of control excursion (hence dynamic range).

3.4 Talbot VCA for High Bandwidth

In February of 1982, Daniel Talbot [15] was issued a USA patent number 4,316,107 for a wide bandwidth VCA. This wide bandwidth was made possible by making the core multiplier circuitry as simple as possible and no feedback amplifier for the log portion of the circuit. An all-diode core is used,

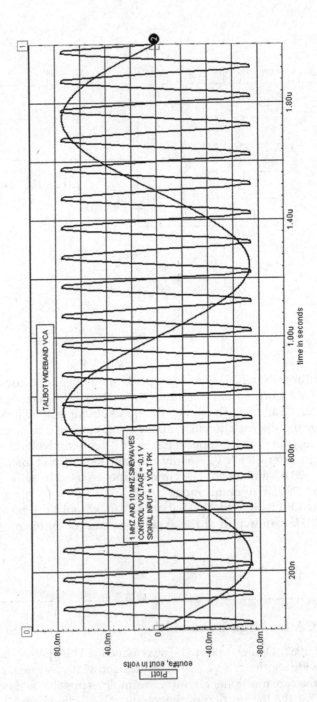

FIGURE 3.5
Talbot VCA at VC = −0.1 volt.

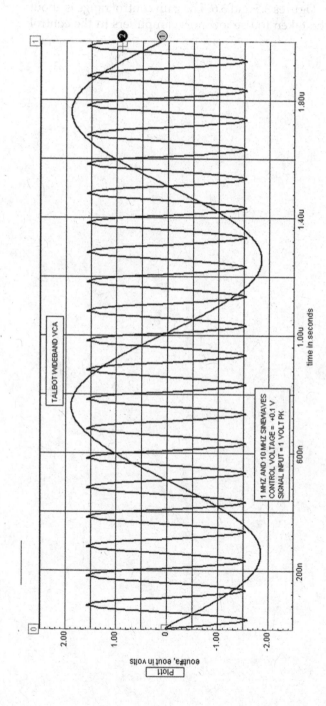

FIGURE 3.6
Talbot VCA at VC = +0.1 volt.

and the diodes can be very high frequency hot-carrier (Schottky) diodes (matched of course). The resulting SPICE model is shown in Figure 3.4 and performance is shown in Figures 3.5 and 3.6. The gain control range is about 30 dB. Great care must be taken to use low-noise amplifiers in the control circuitry.

4

Emitter Followers and Source Followers (FETs)

4.1 Introduction

A problem sometimes surprises the designer of unity gain voltage followers implemented as common-collector or common-drain topologies. The surprise is oscillation. This chapter offers simplified explanations and a solution.

4.2 Model for a Bipolar Junction Transistor (BJT) (Emitter Capacitor Loaded) Simplified

Refer to Figure 4.1.

$$Z_{IN} = \frac{\beta}{j\omega C} \tag{4.1}$$

$$\beta = \left(\frac{f_T}{jf + f_\beta} \right) \tag{4.2}$$

where
  f_t = the frequency in Hz where beta falls to unity
  f = frequency in Hz
  f_b = beta low-frequency corner in Hz

Therefore:

$$Z_{IN} = \left(\frac{f_T}{(jf + f_\beta)(j2\pi fC)} \right) \tag{4.3}$$

DOI: 10.1201/9781003088547-4

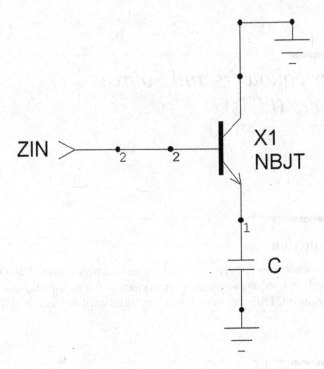

FIGURE 4.1
Model of emitter follower (dc bias is implied).

or:

$$Z_{IN} = \left(\frac{f_T}{C(-2\pi f^2 + j2\pi f_\beta f)} \right) \tag{4.4}$$

which has a negative real term.

4.3 Potential Oscillation in BJT Emitter Follower and Explanation

Note the innocent looking topology of Figure 4.1. It is a model which assumes all direct current (DC) biasing has been accomplished. The transistor BJT emitter-follower model with more detail is shown in Figure 4.2.

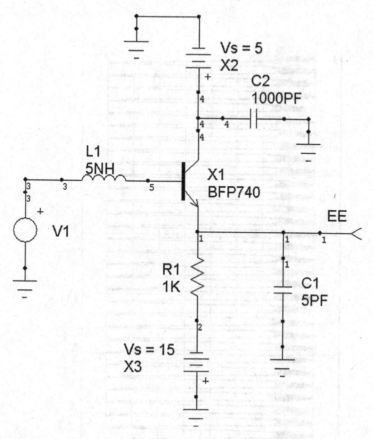

FIGURE 4.2
Oscillating emitter follower.

A negative impedance appears at the input through L1. Since both the load capacitor and the transistor beta are proportional to $\dfrac{k}{jw}$, the product has the form of $\left(\dfrac{k}{j\omega}\right)^2 = \left(\dfrac{k^2}{jj\omega^2}\right) = -\left(\dfrac{k^2}{\omega^2}\right)$.

This negative resistance will find a willing accomplice in stray reactances (L1 and C1) to form an oscillator. The inductance of a short piece of printed circuit board (PCB) track or an integrated circuit (IC) bond wire will represent the 5 nH L1 parasite. C1 has a ballpark value of 5 pF to account for PCB traces, etc. With careful layout this value can be < 2 pF. So, for caution, the circuit designer should routinely add 50 to 100 Ohms in series with each BJT base or FET gate. This will offset the negative resistance at the BJT base or the FET gate, thus preventing unwanted oscillation. See Figure 4.3.

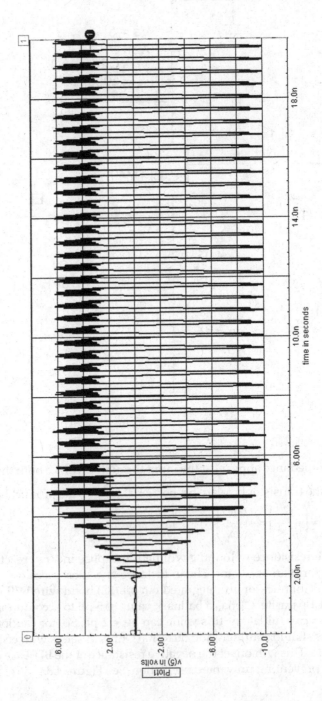

FIGURE 4.3
Waveform at base of BJT shows oscillation.

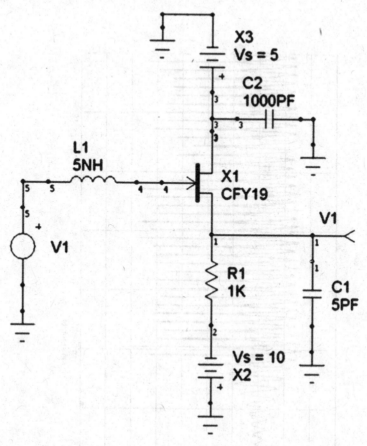

FIGURE 4.4
Oscillating source follower schematic.

4.4 Actual Simulation of Field Effect Transistor Source Follower Showing Oscillation

Figures 4.4 and 4.5 shows FET source follower oscillation due to negative resistance, and Figure 4.6 demonstrates that the oscillation can be killed by placing 50 Ohms in series with the gate of the FET.

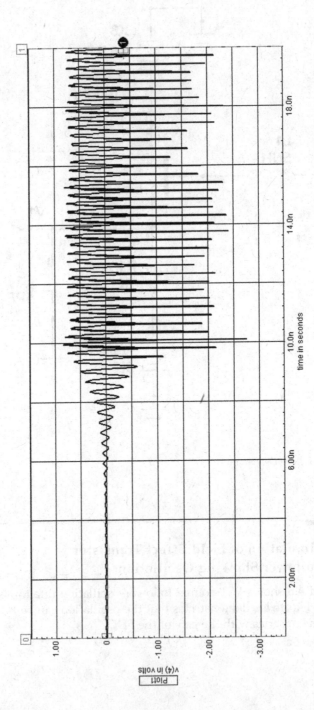

FIGURE 4.5
Oscillating source follower waveform.

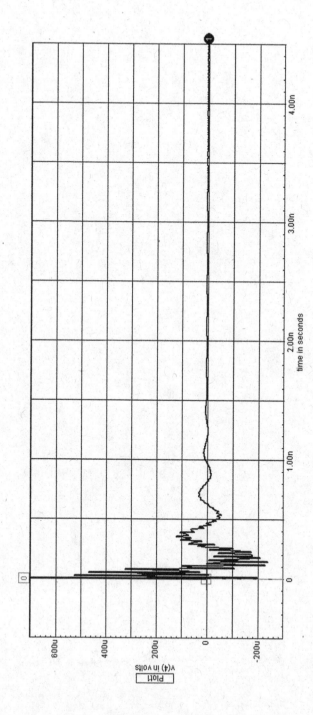

FIGURE 4.6
Non-oscillating source follower after 50 Ohm series with gate.

5

Equally Terminated Two-Port Reciprocal Networks and Reversal of Input and Output

5.1 Introduction

In this chapter, the termination style of passive filters is discussed, specifically double equal termination. It is shown that because of reciprocity and equal source and load impedance, the gain in both directions will be equal (s12 will equal s21).

5.2 What Is Meant by Equally Terminated (Doubly Terminated)

A filter can be singly terminated, which means that a resistive element exists at only one end of the filter (e.g., 50-Ohm source driving the filter and infinite Ohms loading the other end of the filter) or doubly terminated (e.g., 50-Ohm terminations at both the input and output of the filter). A bandpass filter, having its input driven by a finite resistance but its output loaded by infinite resistance, can be designed to have voltage gain; a doubly terminated filter cannot.

5.3 Example of a Reciprocal Two-Port Network Driven by Equal Source and Load Impedance

Refer to the network of Figure 5.1. Its gain in a 50-Ohm source/load environment is shown in Figure 5.2 and 5.3 (notice the curves for s21 and s12 are superimposed on each other).

DOI: 10.1201/9781003088547-5

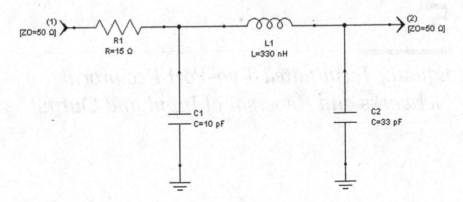

FIGURE 5.1
Schematic of generalized reciprocal network example.

5.4 Simulation of Network s21 and s12 (Gain in Either Direction) Showing s12 = s21

No matter what components are in the filter, its gain (or loss) will be identical in both directions *if the filter is doubly terminated* in the same impedance. That is to say, if we mirror the input and output nodes of the filter, the values of s21 and s12 will be identical (s21 is forward gain in dB and s12 is gain in the reverse direction) in dB.

5.5 Asymmetry of Components Makes s11 ≠ s22 (Example Figure 5.1)

However, in general, the reflection coefficients s11 and s22 do *not* have to be equal (these are usually represented in dB and are also known as return loss (with negative sign). This is especially true in this filter example because the loss is piled up at the input of the filter due to resistor R1 at node 1, so naturally s11 will be less reflective (better return loss) in the deep stop-band (higher frequencies in this example). A "normal" filter accomplishes its rejection by being reflective (either toward a short or an open). There is a class of filter called "constant- impedance" which does not exhibit that property. We will not discuss those filters here.

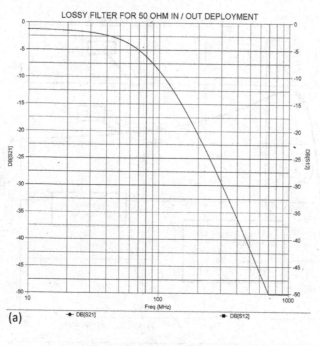

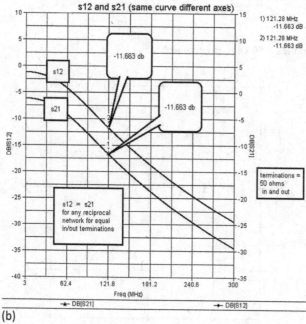

FIGURE 5.2
(a) s21 and s12 of lossy filter example. (b) Graph of s21 and s12 reciprocity in 50-Ohm environment.

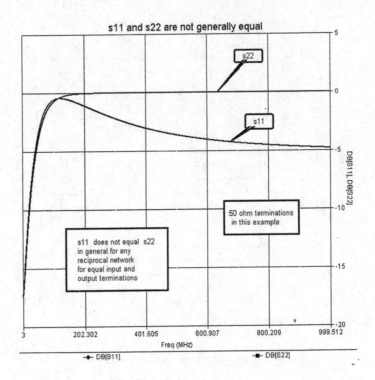

FIGURE 5.3
Graph of s11 and s22 in 50-Ohm environment.

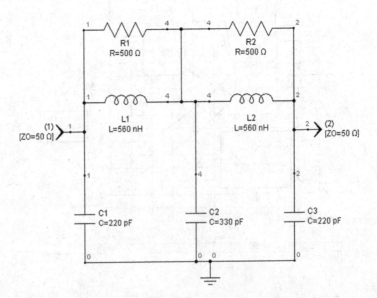

FIGURE 5.4
Symmetrical lowpass filter schematic.

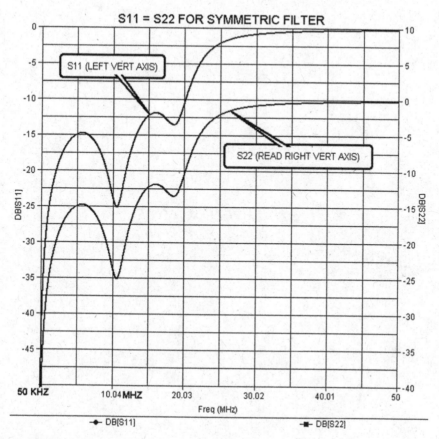

FIGURE 5.5
s11 equals s22 for a symmetric filter (for example, of condition shown in Figure 5.4).

5.6 Symmetry of Components Makes s11 = s22, with Example

In Figure 5.4 a lowpass filter schematic is shown with component mirror symmetry around the midpoint.

Such symmetry results in equality of s11 and s22 as shown in Figure 5.5. If vertical axes were the same, the two curves would superimpose on each other.

6

Importance of Terminating Filters Properly

6.1 Introduction

The importance of correct filter termination impedance is presented with examples. A second order lowpass filter (LPF) is described which exploits the termination impedance to make useful voltage step-up transformers.

6.2 Single Termination of Simplest LC (Inductor-Capacitor) Second Order Lowpass Filter

In Figure 6.1 we see the schematic for a simple network used as either a notch or a lowpass filter, depending on where the output port is. Figure 6.2 shows

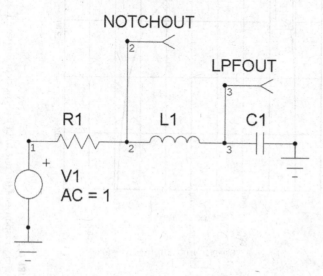

FIGURE 6.1
Simple lowpass or notch network.

DOI: 10.1201/9781003088547-6

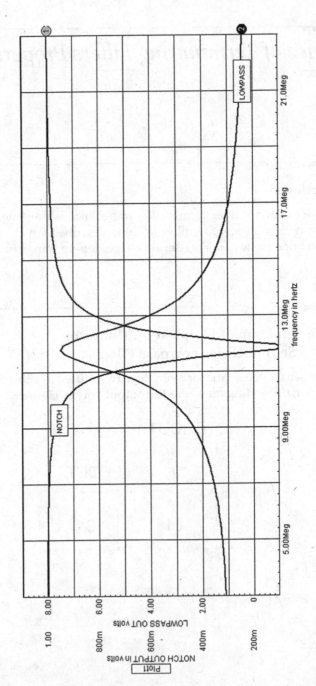

FIGURE 6.2

Frequency response of the notch/LPF network of Figure 6.1.

the frequency response for this example when L1 = 10 microHenries and C1 = 18 picoFarads.

6.2.1 Frequency Response for the Case of Peaking (Voltage Gain before Rolloff)

Although the curve for the lowpass filter resembles a bandpass response near 12 MHz, at frequencies above that it rolls off, eventually transitioning to a stopband. We can reduce how large the peaking is by varying (increasing) the value of R1 (see Figure 6.3).

6.3 Frequency Response for the Case of No Peaking

When the voltage across C1 is about equal to the voltage drop across R1, we have better flatness of frequency response, as can be appreciated from multiplying the current around the loop (V1/R1) by the impedance of C1 at resonance. If R1 = 1000 Ohms, then the loop current is 1 milliamp for a 1-volt source. The impedance of the capacitor C1 at resonance is 750 Ohms (at 11.8 MHz). The amplitude of the output is thus 750/1000 or -2.5 dB (when source is 1 volt).

6.4 Lesson: Even Such a Simple Network Behaves Radically Different for Incorrect Termination

We have just seen the effect of filter termination for a simple filter. This radical sensitivity to termination is evident on most filters. This suggests, for example, that a resistive pad (perhaps 3 or 4 dB) be placed between a filter and the source driving it (e.g., a mixer, whose output port may be a complex impedance or non-standard value). Mixers and filters are adversaries because signal frequencies in the filter stopband present short or open circuits to the mixer, causing spurious signal products, so small values of resistive attenuation offer extreme benefit. A 3-dB pad improves the reflection coefficient by 6 dB (reflected behavior encounters both directions through the pad).

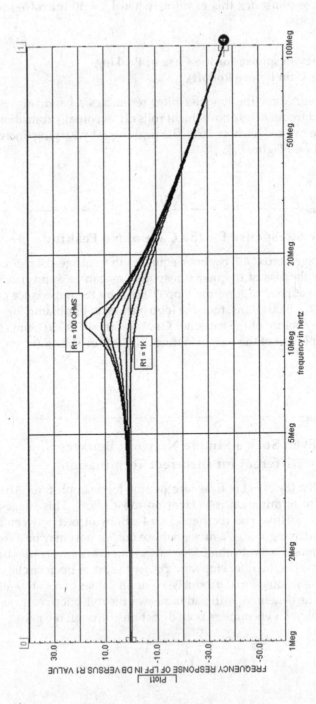

FIGURE 6.3
Frequency response in dB of the LPF network of Figure 6.1 with value of R1 varied (single terminated filter).

6.5 Sometimes This Filter Is Useful for Its Peaking Ability to Make a Narrow Band Transformer

The simple filter in this example can be used to create voltage gain at a single frequency. Consider the case where R1 = 22 Ohms. We would have a transformer-like voltage gain at 11.8 MHz of 20 log (750/22) = 31 dB assuming the load across C1 is infinite.

6.6 An Equally-Terminated (Doubly-Terminated) Filter Can Never Have Voltage Gain

The reason that no filter can have voltage gain if terminated equally at both input and output ports is that it represents the maximum power transfer case without power gain. Any filter ripple peaks must be equal to or less than 0 dB (0 dB already factors in the loss from the equal 50-Ohm terminations). However, for single termination filters we can have voltage gain at the filter ripple peaks.

7
Diode Detector Flatness

7.1 Introduction

In this chapter, the topology for a diode radio frequency (RF) detector is discussed, and the quasi-peak form is the one that works best. A discussion of frequency response flatness versus source resistance is provided.

7.2 Diode Detector Configurations that Do Not Work

It is tempting to configure an RF detector as shown in Figure 7.1, but the result is almost zero output at high frequencies. This is due to the 1 pF shunt capacitance of the diode D1. The waveform at D1's anode is coupled to its cathode enthusiastically by the diode's shunt capacitance, so little or no rectification takes place. The output direct current (DC) versus frequency is plotted in Figure 7.2.

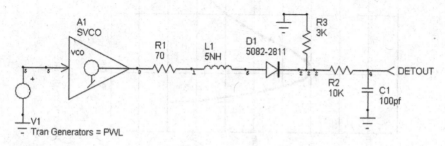

FIGURE 7.1
Averaging RF detector schematic.

DOI: 10.1201/9781003088547-7

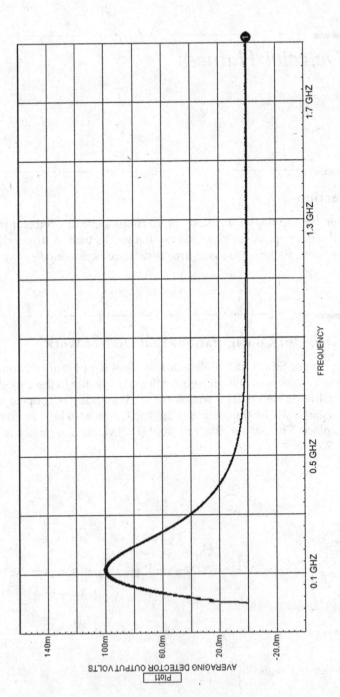

FIGURE 7.2
Averaging RF detector response.

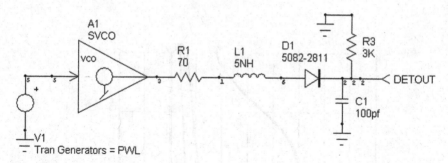

FIGURE 7.3
Quasi-peak detector schematic.

7.3 Peak Detector Configuration Yields the Flattest Response

The correct configuration for RF detection is the quasi-peak detector mode as shown in Figure 7.3, because the coupling across the diode is minimized when not conducting so rectification can occur. The diode's inductance plus printed circuit board (PCB) trace inductance is modeled as L1 (about 5 nH in the example). This inductance drives the diode's shunt capacitance (about 1 pF) to form a second order lowpass filter driven by source resistance R1. Varying the value of R1 will vary the amount of peaking in this lowpass filter equivalent. Some curves for the frequency response of the circuit are shown in Figure 7.4 corresponding to various values for R1 (which includes the source's 50 Ohms). Using R1 = 75 Ohms seems to yield the flattest response to 1 GHz.

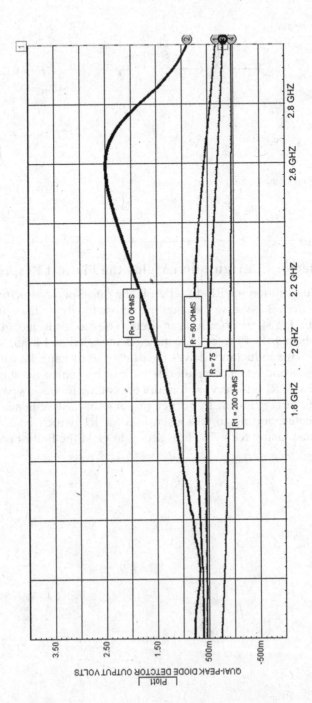

FIGURE 7.4
Quasi-peak RF detector response vs R1.

8

Passive Filters

8.1 Introduction

Passive filters are discussed in this chapter, along with topics including group delay, group delay correction, canonic filter forms, notch networks and a technique to make the notches infinitely deep, scaling filter impedance and operating frequency, and conversion of a lowpass to a bandpass type. The coupled-resonator approach to bandpass filters (BPF) is discussed along with simple impedance matching techniques. Surface acoustic wave (SAW) filters are discussed with remarks on triple transit echo. The chapter ends with tone burst responses to be expected from notch and lowpass filters (LPFs) and a brief discussion of state variable filters and modeling.

8.2 LC Passive Filters

Many types of filter are intended to interface with either a single or double termination resistance [9, 11, 12]. The double resistance terminations do not have to be equal-valued, but it is most common to design for 50-ohm environments at both ports for the following reasons.

a) Some filters require tuning or testing, and the equipment for that purpose features an internal generator and load impedance of 50 Ohms.

b) At higher frequencies, most commercial components use 50-Ohm terminators (amplifiers, mixers, attenuators, etc.).

c) At higher frequencies, the impedances of most circuitry decrease in value making high-impedance designs impractical if not impossible.

LC filters are mainly used in radio frequency (RF) situations, because at audio frequencies they would be physically too large. There are important exceptions, however, such as the crossover networks in loudspeaker systems. We shall focus mainly on RF/IF (intermediate frequency) and video filters.

DOI: 10.1201/9781003088547-8

8.3 Types of Filters: Lowpass, Highpass, Bandpass, Bandstop, and Allpass

From Figure 8.1, it is obvious that bandstop filters (top trace bracketing 9 and 11 MHz) pass low and high frequencies while rejecting middle frequencies, according to a shape dictated by the designer. A highpass filter (bottom trace) does the obvious (i.e., it passes high frequencies). A bandpass filter passes a range of signal frequencies and rejects others below and above that range. A simple notch filter can be realized by subtracting the bandpass response from the unfiltered path. That would make a narrow V-shape notch which barely qualifies as a bandstop filter. We will discuss a better way to make a notch filter in Section 8.12. An allpass filter is one that passes all signals equally regardless of frequency. It is a specially tailored type which modifies the phase versus frequency according to a desired curve, thus attempting to correct the varying phase versus frequency of one of the other filter classes. It is sometimes called a group delay compensator.

In the majority of filters, suppression of unwanted signal frequencies is accomplished by the filter reflecting energy away from itself (100% reflection coefficient or 0 dB value) by becoming a short or an open circuit.

The reflection coefficient is either s11 or s22 (depending on the port of interest), and expressing it in dB implies a scalar value. However, in reality, the scalar can be represented as a vector if a vector network analyzer is used to measure it. Often, the vector network analyzer (VNA) will report the value of s11 or s22 in magnitude and phase, or sometimes in real and imaginary values. This allows better modification to a perfect 50 ohms by adding shunt or series resistance. A BASIC program to convert s11 to z11 is listed below. It is written in BASIC language, and may need adaptation to your operating system version.

```
10 REM    PROGRAM TO COMPUTE ZIN FROM S11
20 REM
30 CLS
40 PRINT "PROGRAM TO COMPUTE ZIN FROM S11"
50 PRINT:PRINT
60 INPUT "S11 MAGNITUDE = ";SMAG
70 INPUT "S11 ANGLE (DEG) =";SDEG
80 INPUT "ZO SYSTEM IMPEDANCE, OHMS =";ZO
90 PRINT
100 GAMMA=SMAG*COS((SDEG/180)*3.14159)
110 GAMIM=SMAG*SIN((SDEG/180)*3.14159)
120 NUMRL=1+GAMMA
130 DENRL=1-GAMMA
140 MAGNUM=SQR(NUMRL*NUMRL+GAMIM*GAMIM)
150 MAGDEN=SQR(DENRL*DENRL+GAMIM*GAMIM)
```

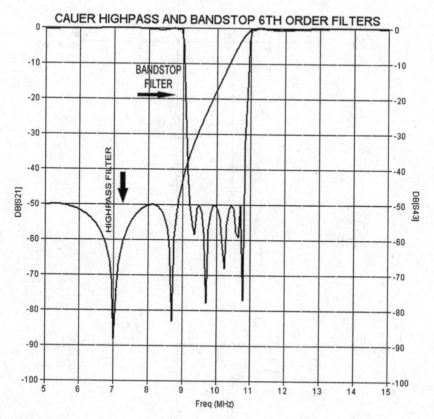

FIGURE 8.1

Bandstop and highpass filter examples.

```
160 ANGNUM=ATN(GAMIM/NUMRL)
170 ANGDEN=ATN(-GAMIM/DENRL)
180 MAGZ=MAGNUM/MAGDEN
190 ANGZ=ANGNUM-ANGDEN
200 PRINT "MAGNITUDE OF ZIN = ";MAGZ*ZO
210 PRINT "ANGLE OF ZIN = ";ANGZ*57.3;" DEGREES"
211 PRINT
212 INPUT "TRY AGAIN (Y/N) ";A$
213 IF A$="Y" THEN GOTO 50
214 IF A$="y" THEN GOTO 50
220 STOP
230 END
```

The benefit of knowing the impedance represented by s11 to ZIN1 conversion is to be able to improve the s11 number at a point in frequency, as shown by schematics Figure 8.2a and b (to the slight detriment of s21) and corresponding graphs of Figure 8.3a and b.

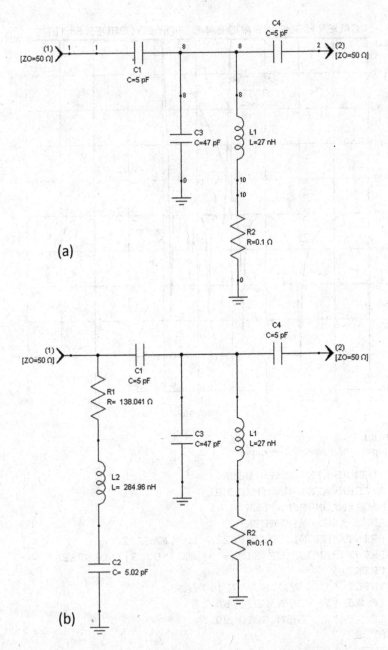

FIGURE 8.2
(a) Second order BPF schematic with no return loss improvement. (b) 2nd order BPF schematic with return loss improvement.

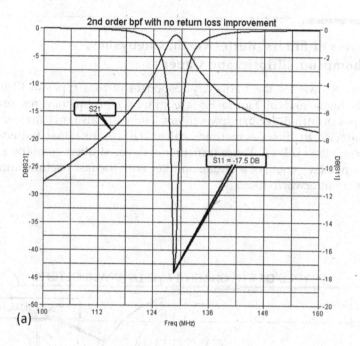

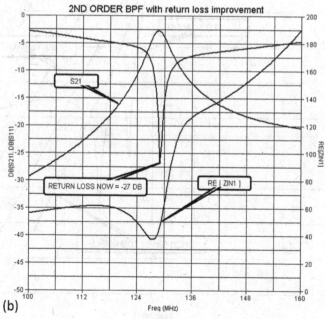

FIGURE 8.3

(a) Second order BPF with no return loss improvement. (b) Second order BPF with return loss improvement.

8.4 Forms of Filters: Butterworth, Chebyshev, Thompson, Elliptic, and Cauer

In Figure 8.4 we see the frequency responses of four types of fifth order lowpass filters and in Figure 8.5 are graphs of frequency responses of four types of fifth order bandpass filters. There are also mixtures of Bessel and Butterworth filters sometimes called TBT (transitional Butterworth–Thomson) filters [23, 24]. Those transitional flavored filters combine some of the group delay benefit of a Thomson filter with the fast risetime and selectivity of a Butterworth filter.

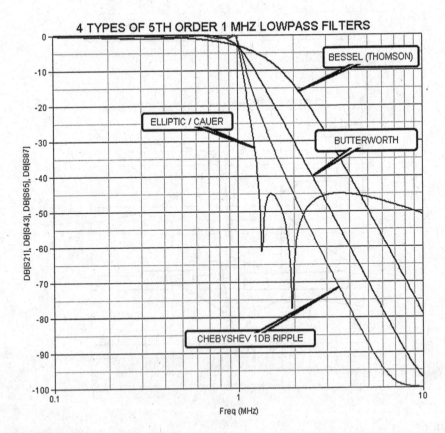

FIGURE 8.4
Four types of 1 MHz fifth order LPFS.

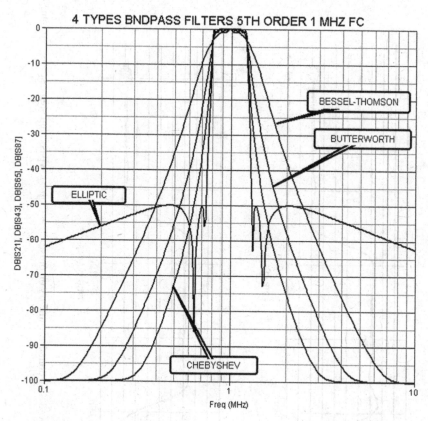

FIGURE 8.5
Four Types of fifth order bandpass filters.

8.5 Group Delay

A very important property of a filter is its group delay curve. In many applications it is important for flatness of delay across most of its passband. Observe the time domain plots of Figure 8.6, the upper plot showing the pulse response of the basic lowpass filter, and the lower trace demonstrating the result of the application of mild group delay equalization. In the upper trace, the step response waveform has no pre-shoot as a pulse occurs, but has overshoot at the final value of the pulse rise time. In the lower plot, notice that the pre-shoot and overshoot on each pulse edge is symmetrical, that is, the "horns" at the pulse edges are similar for each edge. There is a filter class called Bessel which has a gentle rolloff into the stopband, but not much selectivity, whereas a more "brick wall" filter (a rectangular frequency

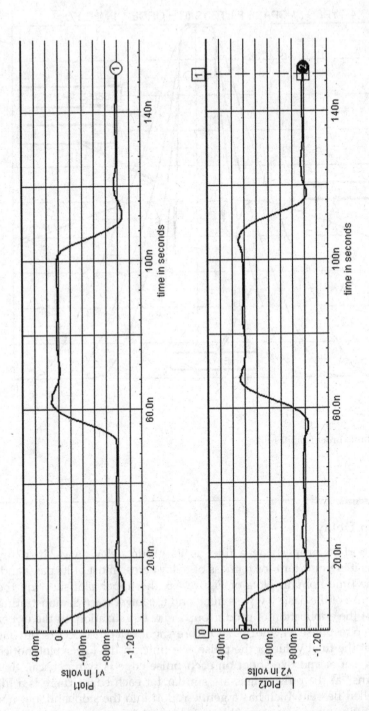

FIGURE 8.6

Before and after group delay compensation demonstrates improved time domain symmetry on pulse edges (Top = lowpass filter step response; bottom = lowpass filter delay equalized step response).

response shape) has plenty of selectivity but poor time domain fidelity. This is because of the Gibbs phenomenon, which suggests that rectangularity in the frequency domain results in sin-X-over-X ringing in the time domain and rectangularity in the time domain causes sin-X-over-X behavior in the frequency domain.

This is illustrated by Figures 8.7, 8.8, and 8.9.

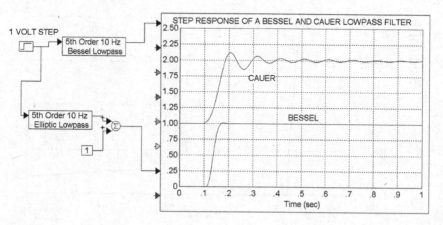

FIGURE 8.7
Step response of a fifth order Bessel and Cauer LPF.

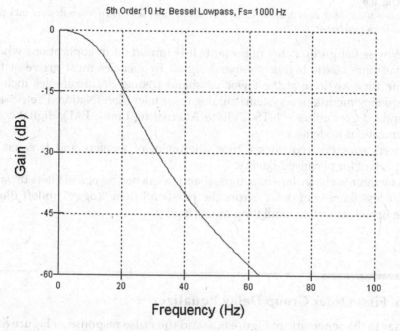

FIGURE 8.8
Frequency response of Bessel LPF shows low selectivity.

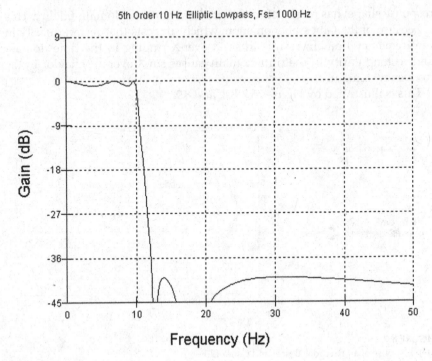

FIGURE 8.9
Frequency response of fifth order Cauer LPF shows high selectivity (steep descent into stop band from pass band).

Why is flat group delay important? It is important in applications where wavefronts of the higher passband signal frequencies must arrive at the same time as those at the lower passband frequencies. Examples include frequency modulation systems, analog color television (National Television Standard Committee – NTSC, Phase Alternating Line – PAL), digital communications systems, etc.

Even at audio frequencies, time alignment of woofers and tweeters is important for perceived fidelity.

The more rectangular-shaped passband frequency response filters usually offer less flatness of delay across the passband than "soggy" rolloff filters like Bessel, as well as exhibiting more ringing.

8.6 First Order Group Delay Equalizer

Refer to the schematic of Figure 8.10 and the pulse response in Figure 8.11. Its frequency response is called "allpass" because it passes all frequency

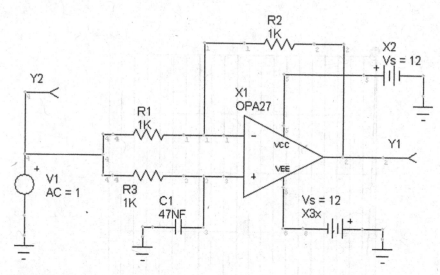

FIGURE 8.10
Schematic of first order active allpass.

signals with equal gain (Figure 8.12) up to circuit limitation. However, its phase response (Figure 8.13) is not flat, and its step response appears to have a "horn" (Figure 8.11 top trace, at the 10 microsecond point in this example) and then settling toward a reverse phase. This is obvious by inspection; when an input step occurs, the capacitor is a short circuit, making the overall circuit gain negative unity. Then, as the capacitor settles toward an open circuit, the gain becomes plus one. The group delay (again Figure 8.13) is merely the derivative of the phase response, and is expressed in seconds (or nanosec or usec, etc.). Notice the shape of the curve; most group delay occurs near zero Hz. Thus, this order of equalizer is most often used in correcting lowpass filter group delay (augmented by one or more second order equalizers).

8.7 Second Order Group Delay Equalizer

The schematic of one possible realization of a second order equalizer is shown in Figure 8.14 and its phase and group delay curves in Figure 8.15. Notice that the peak of delay occurs near the resonant frequency of tank circuit L1/C1. Also notice that the phase across frequency goes from 180 to −180 degrees, twice that of the first order example, having no (L1) inductor in parallel with C1.

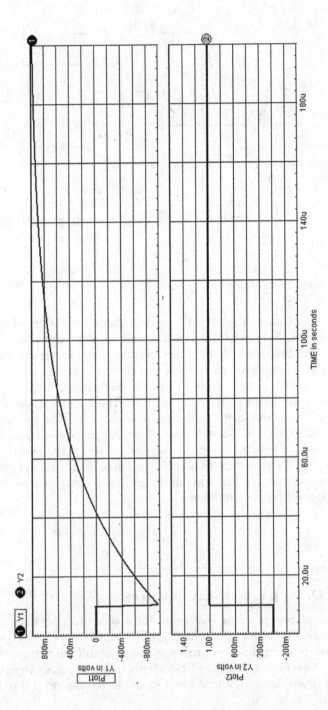

FIGURE 8.11
First order allpass pulse response.

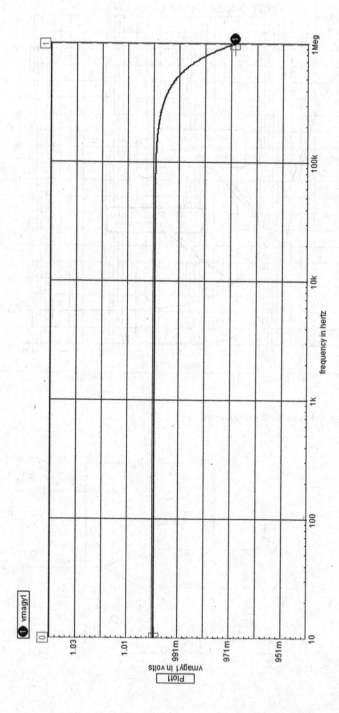

FIGURE 8.12
First order allpass freq. response.

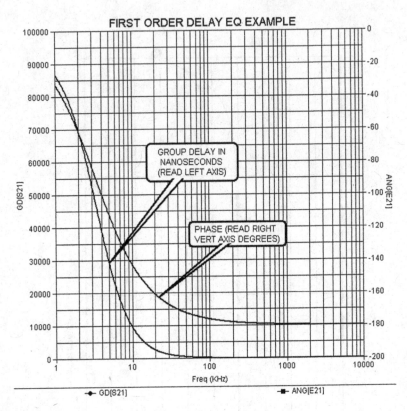

FIGURE 8.13
First order delay EQ phase and group delay.

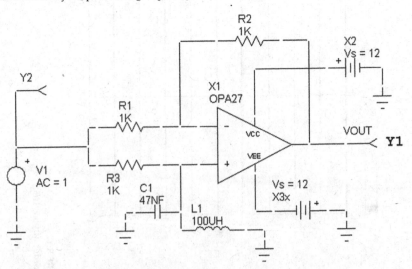

FIGURE 8.14
Schematic of second order active allpass.

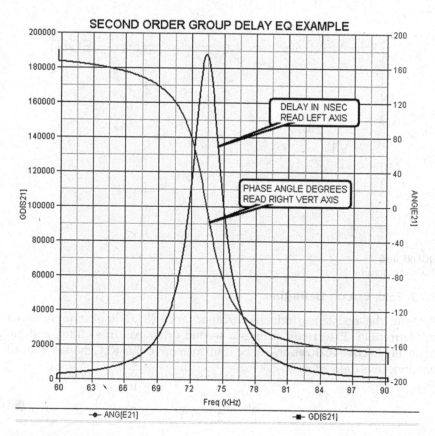

FIGURE 8.15
Second order allpass delay example.

There are many ways to make a second order delay equalizer, some of which use

a) Purely passive networks that offer constant impedance so that they may be cascaded

b) Passive networks that offer compensation of lossy element effects on amplitude flatness, but not necessarily constant impedance

c) Active-passive networks that utilize op-amps and passive components or differential amplification

In awhile we will introduce the (b) approach using familiar components to an RF designer and which offers the best performance up to arbitrarily high frequencies.

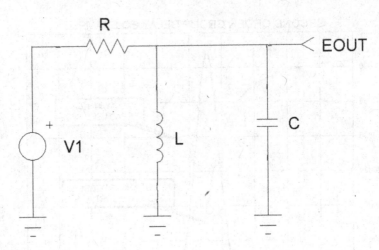

FIGURE 8.16
RLC tank circuit.

8.7.1 Tank Circuit Definitions

Refer to the resistance/inductance/capacitance (RLC) tank circuit shown in Figure 8.16. The circuit defines a second order bandpass filter, and any equivalent active realization can be transformed to this RLC model. The 3dB bandwidth is defined as

$$BW = \frac{1}{2\pi(RC)} \qquad (8.1)$$

where
  BW = 3 dB bandwidth in Hz
  R = tank parallel resistance in Ohms
  C = tank capacitance in Farads
  L = tank inductance in Henries

The tank resonant frequency is given by:

$$f_R = \frac{1}{2\pi\sqrt{LC}} \qquad (8.2)$$

where
  f_R = tank resonant frequency in Hz
  L = tank inductance in Henries
  C = tank capacitance in Farads

Then, we define Q as:

$$Q = \frac{f_R}{BW} \qquad (8.3)$$

Remember that for any ω we have $\omega = 2\pi f$ (i.e., Hz and radians per second are related by 2π).

The phase angle of this tank is given *in radians* as:

$$\theta(\omega) = -2\tan^{-1}\left(\frac{\omega_R \omega}{Q(\omega_R^2 - \omega^2)}\right) \tag{8.4}$$

And the group delay *in seconds* for this second order equalizer stage is:

$$GD = -\left(\frac{d\theta(\omega)}{d\omega}\right) \tag{8.5}$$

Or by taking that derivative,

$$GD = \left(\frac{2Q\omega_R(\omega_R^2 + \omega^2)}{Q^2(\omega_R^2 - \omega^2) + \omega_R^2\omega^2}\right) \tag{8.6}$$

So, for Figure 8.16 to become part of a second order allpass group delay equalizer with L = 100nH, and C = 100pF, and R = 1K ohms, the circuit must have the form of [(2 EOUT) – V1]. The actual allpass circuit is given in Figure 8.17 and the resultant phase and delay is graphed in Figure 8.18.

We check our equation next (it must agree with the simulated answers). For our example, using the values for *R*, *L*, and *C*,

$$Q = 50.4e6 / 1.59e6 = 31.7 \text{ and } \omega_R = 2\pi (50.3e6) = 3.16e8$$

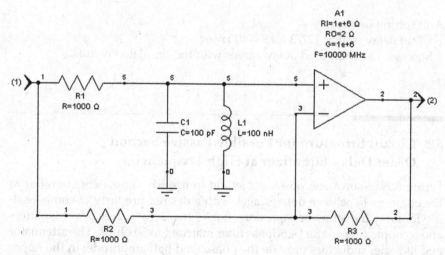

FIGURE 8.17
RLC tank circuit based second order delay EQ.

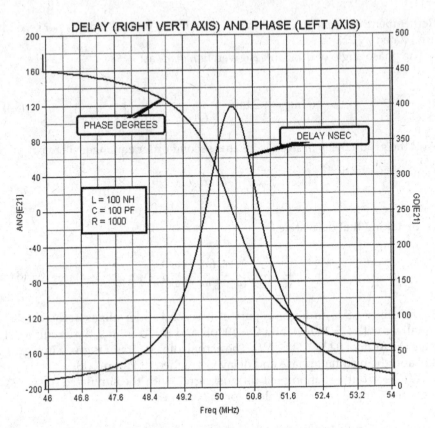

FIGURE 8.18
Actual delay for the equalizer stage values shown.

And setting $\omega = \omega_R$
  GD at delay peak = 127/3.16e8 = 400 nsec
  Success! The computed delay agrees with the simulated result.

8.8 Circuit Structure for Possible Passive Second Order Delay Equalizer at High Frequencies

Figure 8.19a shows one possible method to use RF components familiar to the engineer to achieve delay stages with a desired property of compensating for lossy inductors or capacitors. It uses two zero-degree power splitters and a coupled-resonator bandpass filter matched to 50 ohms. The attenuator and its series inductors provide the phase (and half amplitude) in the upper branch to match that of the lower branch containing the bandpass filter after

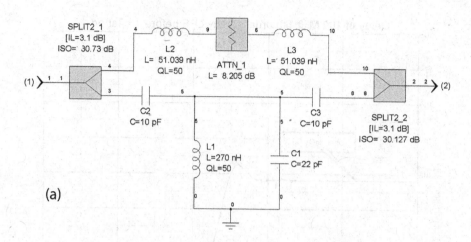

(a)

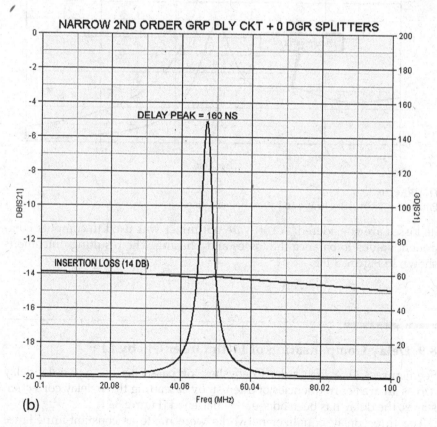

(b)

FIGURE 8.19

(a) Possible realization of a second order delay EQ stage corresponding second order delay and loss curves.

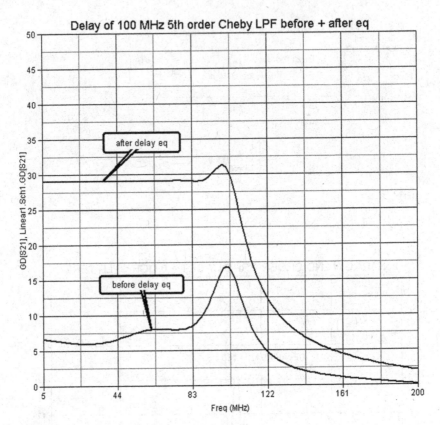

FIGURE 8.20
Fifth order cheby delay before and after compensation.

its losses are considered. A software optimizer was used to compute component values to balance the losses and phasing. The resulting outcome is shown in Figure 8.19b.

8.9 Delay Compensation of Fifth Order Cheby LPF

See Figure 8.20. A fifth order Chebyshev lowpass filter has a non-flat delay across an area of frequencies of interest. By cascading three delay correction stages, the delay has been adequately flattened (Figure 8.21).

The three delay equalizer networks were made as constant-impedance networks (Figure 8.22) so they could be cascaded but we can realize the stages any way we like using active circuitry or well-buffered passive non-constant-Z stages.

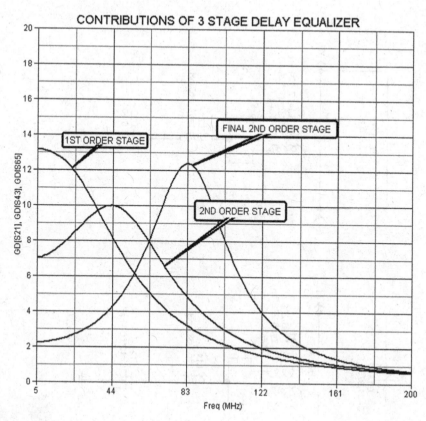

FIGURE 8.21
Fifth order cheby delay equalizer stages.

8.10 First Order Group Delay Compensator

We have previously shown a circuit capable of first order delay equalization. Setting

$$a = \frac{1}{RC} \qquad (8.7)$$

we get the group delay as:

$$GD = \left(\frac{2a}{a^2 + \omega^2} \right) \qquad (8.8)$$

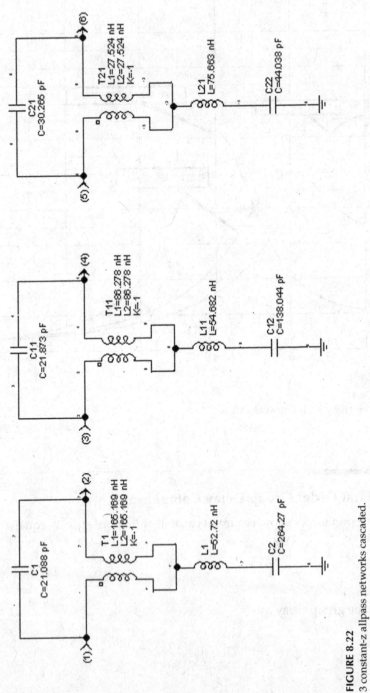

FIGURE 8.22
3 constant-z allpass networks cascaded.

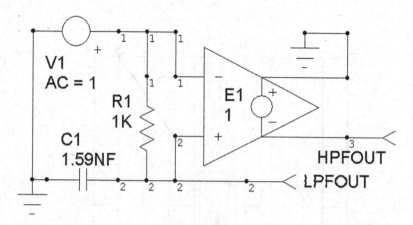

FIGURE 8.23
6 DB per oct HPF by subtracting 6 DB per oct LPF.

8.11 Filters Derived by Subtracting Other Filters

There are occasions when we want a free filter derived from subtracting the response of an already present filter. For example, suppose we want a highpass filter (HPF) by subtracting a lowpass filter's output from the unfiltered path. Look at Figure 8.23. A simple RC lowpass filter response can be subtracted from the flat stimulus to yield the highpass filter indicated in the figure. In Figure 8.24 we see the resulting "derived highpass" response. The question arises "can we make a steeper highpass filter by subtracting a steeper lowpass filter?" So, to answer the question we replace the lowpass filter formed by R1 and C1 with a very steep "brickwall" filter (one having much greater than − 6 dB/octave rolloff). The answer is seen in Figure 8.25, and it is *no*. This is due to the lowpass filter having vector, not merely scalar properties. The response is approaching 6 dB/octave at very low frequencies but it suffers a 6 dB peak near the transition region.

8.12 Notch Networks (Traps) with Infinite Depth

Refer to Figure 8.26. If inductor L1 has a series loss resistance R2, then R_null of 4 times R2 will balance the bridge at resonance, yielding infinite notch depth. (Figure 8.27). Figure 8.28 illustrates a higher level of sophistication, and the notch components are in series with the signal path. The inductor L1

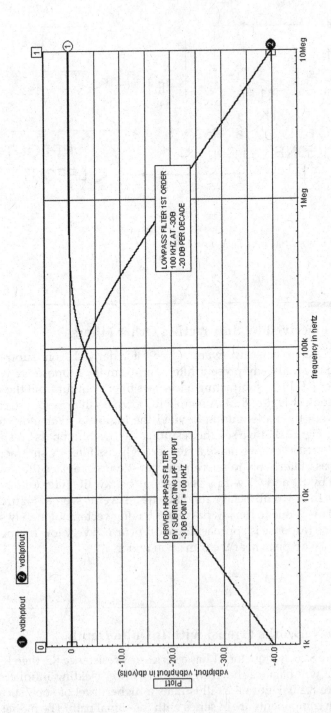

FIGURE 8.24
Derived highpass filter by subtracting LPF.

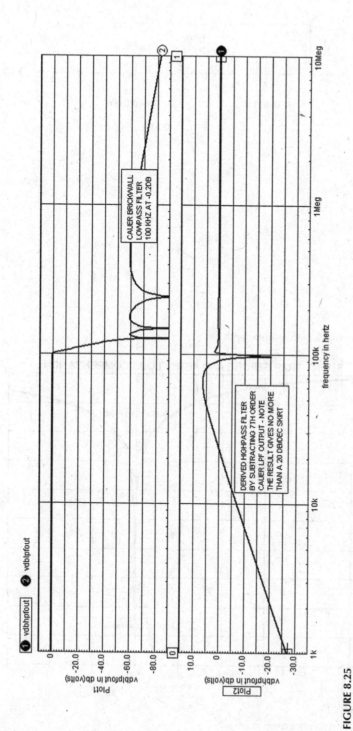

FIGURE 8.25

Failed derived highpass filter by subtracting LPF.

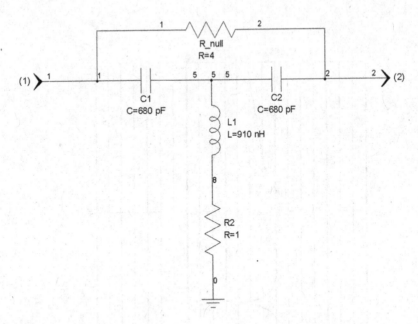

FIGURE 8.26
Shunt notch network with infinite depth.

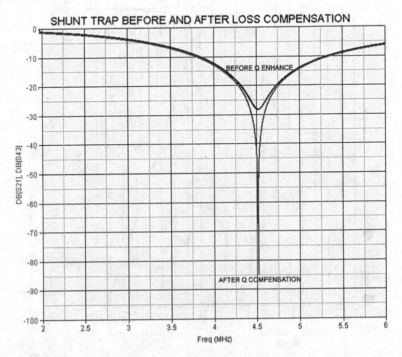

FIGURE 8.27
Infinite notch depth after compensation.

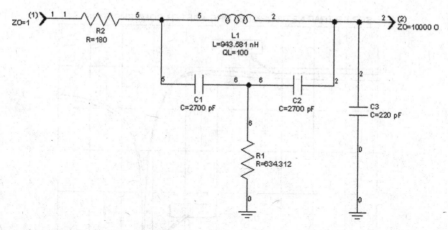

FIGURE 8.28
Series notch network and Q compensation.

has a Q of 100 at 4.5 MHz, modeled as a parallel parasitic resistance of 4 times R1 to balance the circuit and yield an infinite notch depth. It also features an amplitude lift just prior to the notch frequency. This is due to the parallel resonance frequency of L1 and C1 in series with C2 being higher than the peaking frequency, so that at the peaking frequency this parallel tank appears inductive. The resulting network then is equivalent to the model for a lowpass notch network previously discussed. Figure 8.29 plots the frequency response before and after the network notch depth is nulled.

Figure 8.30 illustrates a better network in some situations where we can't easily obtain a sufficiently large value for the inductor. By adding capacitance in parallel with the inductor, below the resulting tank resonance frequency the inductor will appear larger. Benefit is shown in Figure 8.31. This circuit also features an inductor L1 that can be grounded.

8.13 Transforming a Lowpass Filter into a Bandpass Filter

If a lowpass filter can be modeled similar to that of Figure 8.32 with frequency response as in Figure 8.33, it can be transformed into a bandpass filter, having the same terminating impedances by connecting a capacitor in series with each inductor and an inductor in parallel with each capacitor of the lowpass prototype as per Figure 8.34. If we pick 100 MHz as the center frequency of our bandpass filter, each added component must resonate at 100 MHz with its partnered component. This results in the circuit of Figure 8.34 and the response of Figure 8.35 if components were ideal. The resulting

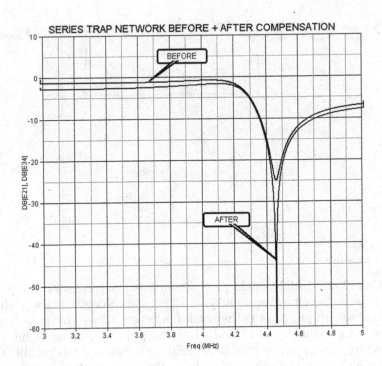

FIGURE 8.29
Series trap before and after compensation.

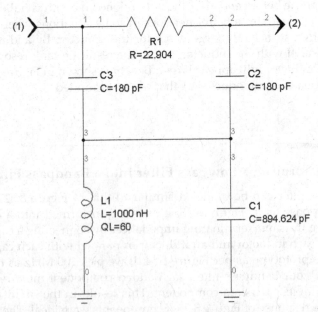

FIGURE 8.30
Grounded inductor shunt trap.

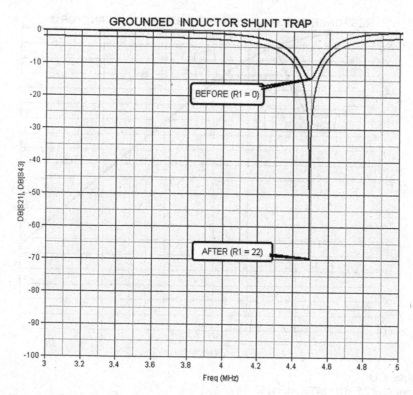

FIGURE 8.31
Grounded inductor shunt trap before and after compensation.

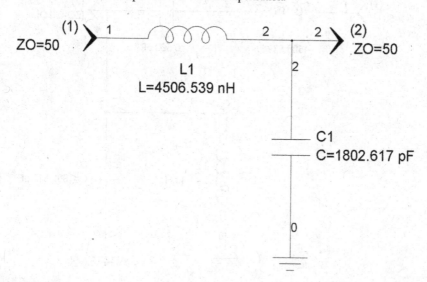

FIGURE 8.32
Schematic of LPF before BPF transform.

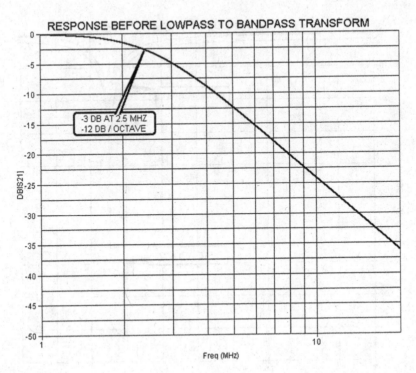

FIGURE 8.33
Response of 2.5 MHz LPF prototype.

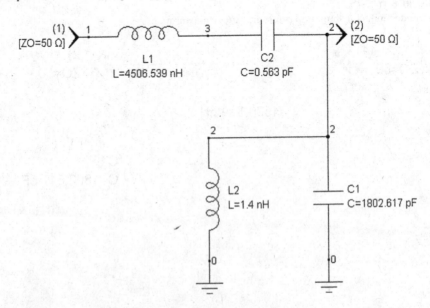

FIGURE 8.34
Schematic after LPF to BPF XFORM.

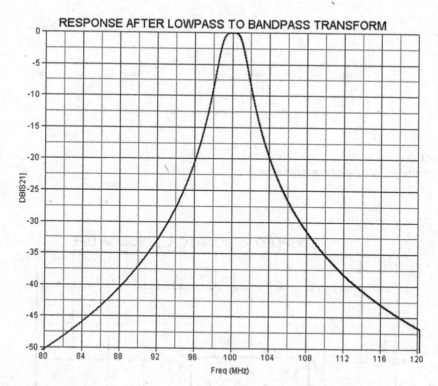

FIGURE 8.35
Response of BPF after LPF XFORM ideal components.

component values, however, are ridiculous. One cannot realize the filter in this form with 50-ohm terminations because:

a) Added inductor L2 at 1.4 nH is too small at 100 MHz and its value would probably be less than the PC board trace connecting it and capacitor C1 is too large and may have a significant parasitic series inductance at 100 MHz.

b) L1 probably has enough parasitic shunt capacitance to become resonant below 100 MHz.

c) C2 (0.563 pF) being so small in value forces L1 to be too large.

There is a much smarter strategy shown in Figure 8.36, called the *coupled-resonator* approach and it is obvious why from the schematic. The resulting response is shown in Figure 8.37. The coupled-resonator network approach is well covered in the literature [12]. For this example, we picked 1500 ohm terminators. Matching to 50 ohms is simple and will be covered in Section 8.16.

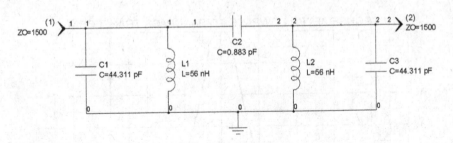

FIGURE 8.36
BPF after using coupled-resonator approach.

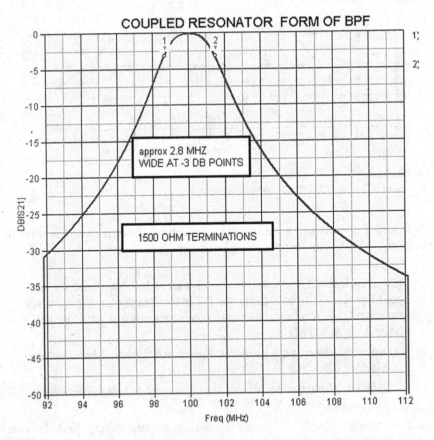

FIGURE 8.37
Response of coupled-resonator example.

8.14 Impedance Scaling a Filter

Most filters' component values can be scaled to yield a new lower filter termination impedance by simply scaling all inductor values downward and all capacitor values upward. Thus, if we want our 1500 Ohm filter to be a 1000 Ohm filter, we would multiply all capacitor values by 15/10 and all inductor values by 10/15. The center frequency (or cutoff frequency) will not be affected.

8.15 Frequency Scaling a Filter

To move the center (or cutoff) frequency to a new value, scale all component values by the ratio of the old frequency to the new frequency. Thus, to modify a 100 MHz filter to become a 200 MHz filter, multiply all component values by 0.5, making all inductors and capacitors half of their old values. (44 pF becomes 22 pF and 56 nH becomes 28 nH).

8.16 Simple Method of Impedance Matching

Use the bandpass filter of Figure 8.36 as an example and match it to a 50 ohm system. To do this, we exploit the equivalent circuits at a particular frequency shown in Figure 8.38. If using a series capacitor to accomplish this,

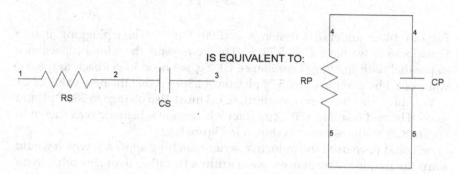

FIGURE 8.38
Series and parallel equivalents at one frequency.

follow the following step-by-step procedure. First, calculate series matching capacitor:

$$C_S = \left(\frac{1}{2\pi f_o \left(R_S \sqrt{\frac{R_P}{R_S} - 1} \right)} \right)$$

(8.9)

where

C_s = the capacitor value in series with the source resistance in Farads
R_s = the new source impedance we want in Ohms
R_p = the old parallel terminating resistance in Ohms
f_o = the center frequency of the bandpass filter in Hertz

Let's plug in some numbers for the 100 MHz coupled-resonator bandpass filter of Figure 8.36. The result is that if we need R_s = 50 Ohms, we have C_s = 5.91 pF. Next we need to know the value of C_p that the matching network provides due to the series to parallel conversion (not the value of C1 or C3).

$$C_P = \left(\frac{C_S}{\left[1 + \left(\frac{R_S}{X_{CS}} \right)^2 \right]} \right)$$

(8.10)

where

$$X_{CS} = \left(\frac{1}{2\pi f_o C_S} \right)$$

(8.11)

For our filter under discussion, X_{CS} = 290 Ohms. Then plugging it into Equation 8.10 we have $C_P = 5.73 pF$. This represents the added capacitance in parallel with the filter capacitance C1. So, we need to get back to the old value of C1 by peeling away 5.73 pF from 44.3pF so that the new value of C1 is 38.57 pF. Our filter is symmetrical, so C3 must also change to 38.57 pF and so on. The end result is a 50-Ohm filter whose new schematic is as shown in Figure 8.39, with response as shown in Figure 8.40.

We could have used the inductive series matching approach which would warp the frequency response toward arithmetic rather than geometric symmetry. The upper skirt will be steeper, in contrast with the capacitive matcher which makes the lower skirt steeper. In both approaches, however, the simple one-component matcher (at either or both ends of the filter) technique is

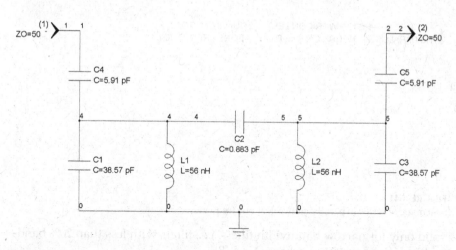

FIGURE 8.39
Schematic of BPF after 50 ohm match via C4 and C5.

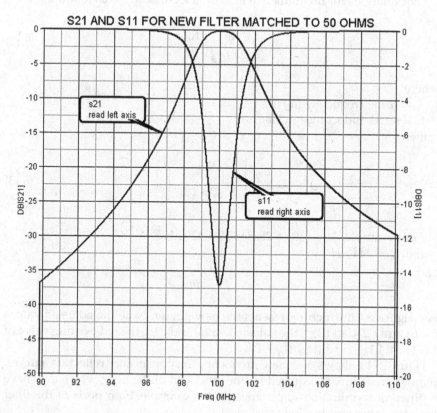

FIGURE 8.40
100 MHz BPF after matching to 50 ohms.

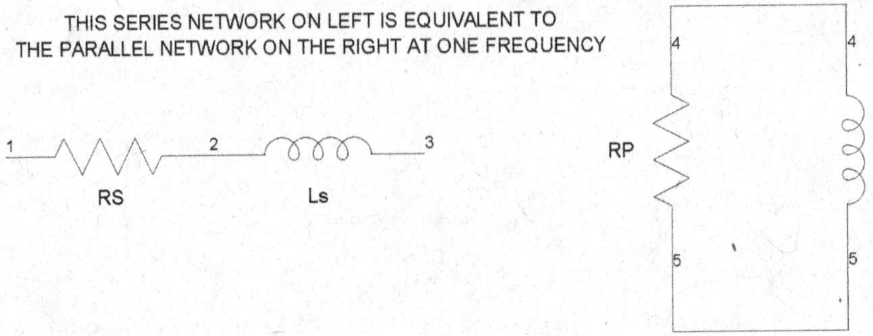

THIS SERIES NETWORK ON LEFT IS EQUIVALENT TO
THE PARALLEL NETWORK ON THE RIGHT AT ONE FREQUENCY

FIGURE 8.41
Schematic of series to parallel network at one frequency.

valid only for narrow bandwidth filters. i.e., filters with less than 20% bandwidth, defined as $100/Q$, where Q is f_o/BW.

Refer to Figure 8.41 and note the equivalence of the series and parallel networks at one frequency.

The equations for the inductive matching technique are as follows:

$$L_S = \left(\frac{R_S}{2\pi f_O} \right) \sqrt{\frac{R_P}{R_S} - 1} \tag{8.12}$$

where
  f_O = *center* frequency in Hz
  L_S = series inductance in Henries
  then

$$L_P = L_S \left[1 + \left(\frac{R_S}{2\pi f_O L_S} \right)^2 \right] \tag{8.13}$$

For our example, $L_S = (50 / 6.28e8) 5.39 = 429 \text{ nH}$
  and $L_P = 444 \ nH$

$$R_S = 50 \ Ohms$$

See Figure 8.42, which represents the new circuit. Note the adjusted values for L1 and L2. Call $L1_{new}$ the adjusted value and $L1_{old}$ the old existing value of L1. Then, $L1_{new} = L1_{old} L_P/(L_P - L1_{old})$.

Figure 8.43 shows the new frequency response and reflection curves. Similar procedure is required for the other end of the filter, which may have a different termination impedance (in the example, both ports of the filter have the same termination resistance).

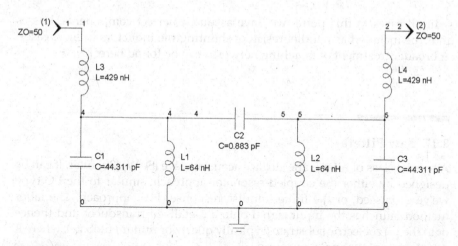

FIGURE 8.42
Schematic of new inductively matched network at one frequency.

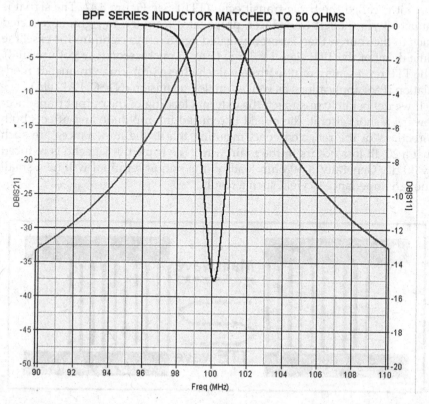

FIGURE 8.43
New inductively matched network frequency response.

In all cases to this point, we have assumed perfect components. Losses must be included as parallel resistors shunting the inductors in most cases. A broader treatment of matching networks can be found here: [39].

8.17 Saw Filters

A special class of filter is the surface acoustic wave (SAW) [30] filter. It can be designed by either the coupled-resonator approach, similar to the LC type we've discussed, or the finite impulse response (FIR) approach. The latter method launches the input signal onto a metalized transducer and thence onto the piezoelectric substrate (typically quartz or lithium niobate), where it now travels at the speed of sound.

The signal is launched and recovered by the transducers as an electric field (voltage, not current) and therefore it can reflect back from the output to the input, where it is re-propagated and because it travels three times back and forth it is called the triple-transit echo (TTE). See Figure 8.44. The signal is launched via the left transducer, propagates from left to right, is reflected back to the source, then is re-propagated again left to right. In this case, the time delay of the TTE signal is large (5 to 15 microseconds is not unusual). The TTE represents a multipath signal which is a problem in frequency modulation (FM) applications as well as analog television (NTSC, PAL, etc.).

If we deliberately mismatch the external loading of the output transducer toward a short circuit, the e-field is reduced in amplitude and affects both directions of the reflected echo, meaning that if the impedances are such that a 20 dB loss occurs in the main wave, the triple transit echo is reduced by 40 dB. Generally, a 50 ohm load on the output terminal will be a small enough impedance to yield such a result.

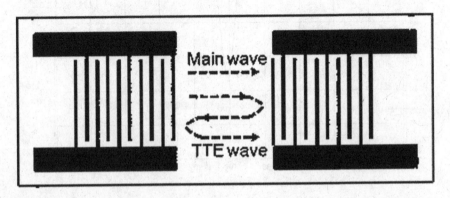

FIGURE 8.44
Triple-transit echo (tte) in saw filters.

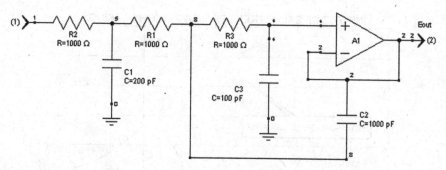

FIGURE 8.45
Third order Sallen-Key-inspired LPF with impedance dip schematic.

8.18 Sallen-Key Inspired Third Order Filters

Among active filters, the Sallen-Key and its cousins are very popular. A third order lowpass structure is shown in Figure 8.45.

As shown in Figure 8.46, it has a negative input impedance component that causes the input impedance to dip in the passband. Figure 8.47 shows how the dip can be eliminated plus other benefits. Finally, Figure 8.48 illustrates the impedance and frequency response of the 3-op-amp form for this filter.

To make a highpass filter, we simply interchange the capacitors and resistors and compute their values. See Figures 8.49 and 8.50.

8.19 Tone Burst Response of a Notch Network or LPF

A notch network or a LPF offers suppression in its steady-state stopband. This does not mean that it can immediately suppress a signal whose frequency falls in that stopband. It takes time for the filter to recognize that it has been delivered a signal that it should suppress. See Figure 8.51. A 10 Hz sinewave has been pulsed on and off (top trace) and then fed to a 10 Hz notch filter and also to a 7.75 Hz LPF. The middle trace shows the response of the notch to this pulsed signal (called a tone burst) and the bottom trace shows the reaction of the LPF to this tone burst.

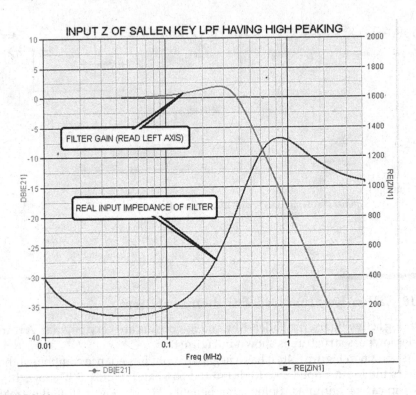

FIGURE 8.46
Third order Sallen-Key-inspired LPF with impedance dip.

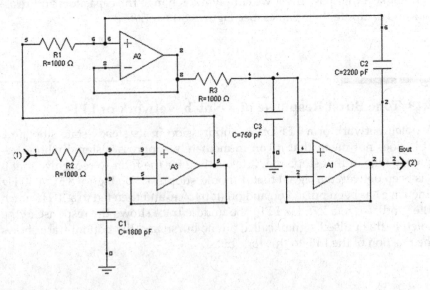

FIGURE 8.47
Third order LPF with no impedance dip schematic.

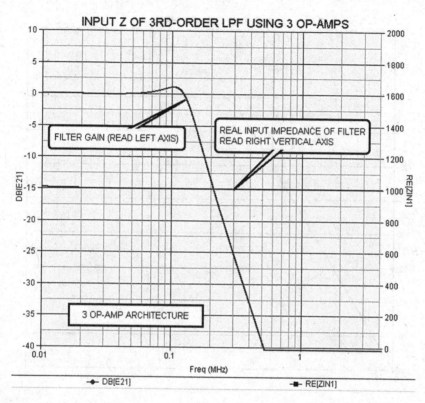

FIGURE 8.48
Third order LPF with no impedance dip.

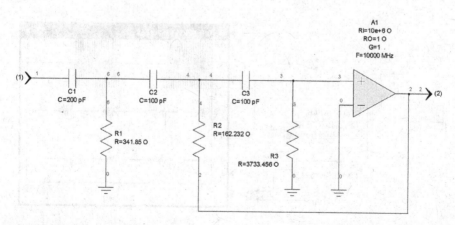

FIGURE 8.49
Schematic of third order voltage follower based HPF.

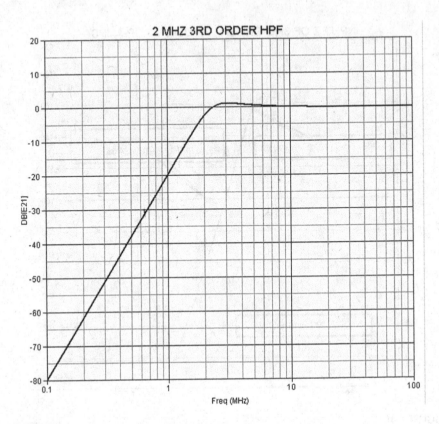

FIGURE 8.50
Graph of third order voltage follower based HPF.

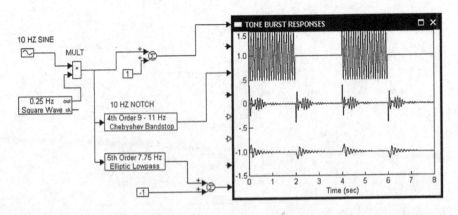

FIGURE 8.51
Tone burst responses of notch or LPF.

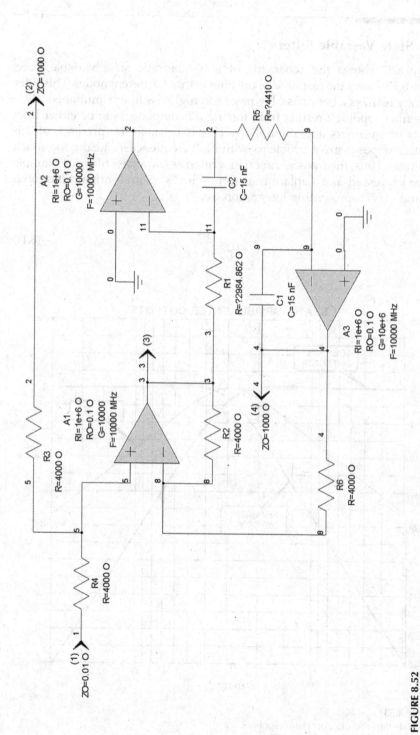

FIGURE 8.52
HKN (Kerwin-Huelsman-Newcomb) state variable filter outputs schematic.

8.20 State Variable Filters

Figure 8.52 shows the schematic of a bi-quadratic state variable filter. Figure 8.53 shows the response of the filter at three different nodes. This filter topology realizes a bandpass, lowpass, and highpass filter simultaneously.

The filter topology results from forcing all components to be either integrators or summers and multipliers. This is because the presence of differentiators poses great problems with high frequencies causing too much amplitude. Thus, the transfer function written as a lowpass filter for example can be expressed as a Laplace transform with "s" representing differentiation and "1/s" representing integration. So:

$$\frac{Y(s)}{X(s)} = \frac{K}{\left(s^2 + as + b\right)} \tag{8.14}$$

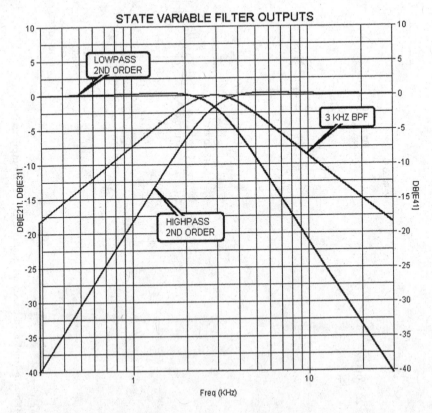

FIGURE 8.53
HKN state variable filter outputs responses.

Rearranging numerator and denominator and dividing both sides by s^2 we thus obtain

$$\frac{KX(s)}{s^2} = Y(s)\left(1 + \frac{a}{s} + \frac{b}{s^2}\right) \tag{8.15}$$

which can be realized using two integrators, plus properly scaled summers and coefficients. The circuit of Figure 8.53 is more complex since it has some numerator terms that are not simple scalar terms, rather the form of the numerator is a quadratic; hence the name bi-quadratic is given to such a transfer function.

9

Secant Waveform for Synchronous Demodulation

9.1 Introduction

This chapter deals with the benefits of the secant waveform to be used as a local oscillator signal to demodulate or frequency translate a modulated carrier to obtain a purer recovered modulation waveform with fewer artifacts.

9.2 Conventional Use of the Cosine Waveform for Synchronous Demodulation

Historically, synchronous demodulation (demod) has employed the cosine waveform (or a square wave) as the local oscillator signal. Refer to Figure 9.1 (top circuit). A modulation (baseband) signal is applied to a mixer (or 4quadrant multiplier) to become frequency translated and centered around $\cos(\omega_c t)$ where ω_c is the carrier frequency in radians per second. Mathematically the transmitted signal can be represented as:

$$S(t) = f(t)\cos(\omega_c t) \tag{9.1}$$

where $f(t)$ = modulation or baseband signal.

Standard recovery method for retrieving $f(t)$ after multiplication by $\cos(\omega_c t)$ is to multiply $S(t)$ by yet another waveform exactly like $\cos(\omega_c t)$ to produce:

$$R(t) = f(t)\cos^2(\omega_c t) \tag{9.2}$$

which can be written as:

$$R(t) = f(t)\left(\frac{1}{2} + \frac{1}{2}\cos(2\omega_c t)\right) \tag{9.3}$$

DOI: 10.1201/9781003088547-9

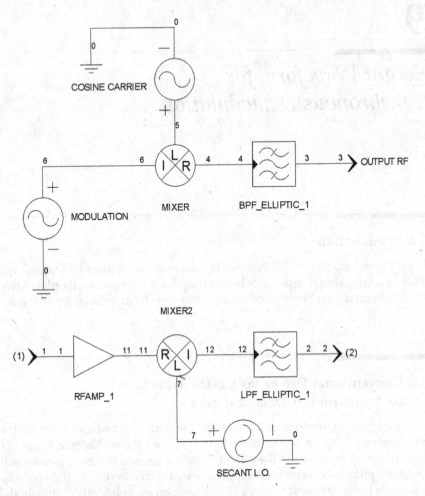

FIGURE 9.1
Standard modulation (top) and secant demod (bottom).

and if $f(t)$ contains frequencies approaching ω_c, there will be images near ω_c. See Figure 9.2 where an image exists slightly above ω_c. There will also be the possibility of the second harmonic of ω_c if $f(t)$ contains any direct current (DC) component (as in the example of Figure 9.2).

9.3 Secant Waveform for Local Oscillator

Let us remember that the mathematically correct way to recover $f(t)$ after demodulation is to multiply the received signal $S(t)$ with the inverse of

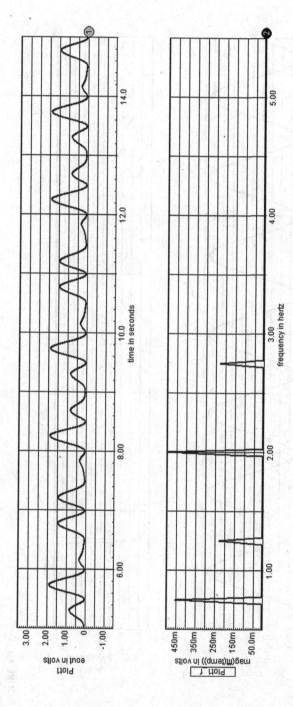

FIGURE 9.2
Conventional demod fast Fourier transform (FFT).

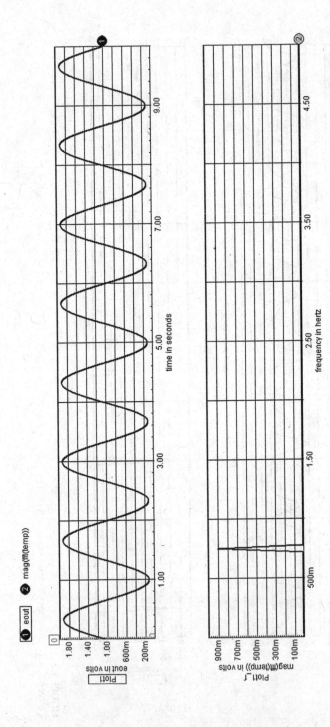

FIGURE 9.3
Ideal secant demod FTT.

the cosine wave, which is a secant wave. In the past this has been difficult because the secant wave has a high crest factor (see Figure 9.4 which has a 15:1 crest factor). But linear multipliers exist (for mixers) that can appreciate the secant waveform with limited crest factor. And the secant wave can be generated via a lookup table. Digital to analog converters exist also to interface the lookup table to the multiplier (not a conventional mixer where the local oscillator (LO) port is operated in saturation).

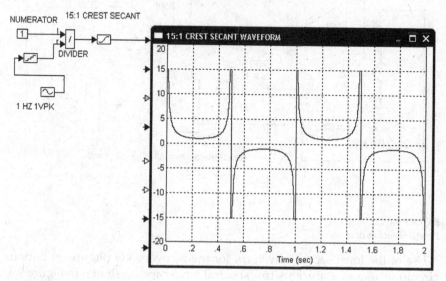

FIGURE 9.4
15:1 crest factor secant waveform.

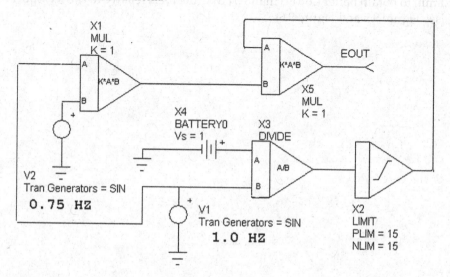

FIGURE 9.5
Secant demod realizes −34 dBc spurious.

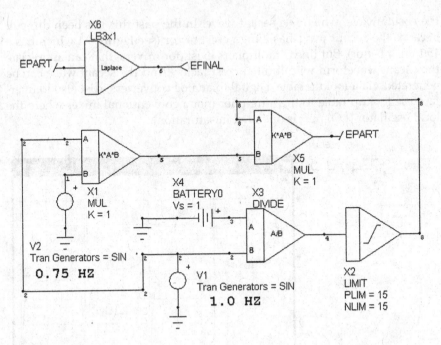

FIGURE 9.6
Practical secant demod.

Use of the ideal secant waveform for the recovery LO (Figure 9.1 bottom circuit) results in a spurious-free spectral landscape, as shown in Figure 9.3. Practically, however, we can use a 15:1 secant crest factor (or up to a practical limit) to obtain better (lower) than −34 dBc (deciBels relative to carrier) spurs (see Figure 9.5 and Figure 9.6).

10

Receiving NRZ Data Using AC Coupling

10.1 Introduction

This chapter presents two methods which can be used to recover transmitted non-return-to-zero (NRZ) data when alternate current (AC) coupling must be employed in the receiver. Such methods are pulse edge differentiation followed by Schmitt triggering or pulse delay line and subtraction followed by Schmitt triggering. The latter method is shown to be superior.

10.2 Edge Detection

Refer to Figure 10.1 in which two cases are shown: in the upper plots the performance of a NRZ ideal transmission and recovery using derivative of the AC-coupled waveform, and in the lower plots recovery is accomplished using a delay line and differencer. In all cases for illustration, a 1 bit/second NRZ code is transmitted as a 1-0-1-0-1-0-1-0 (etc.) pattern (a square wave) and the middle traces show the recovered data waveform. In the upper plot family, the bottom traces show differentiation and delay line method, respectively, of obtaining edge stripping. In the case of the differentiator method, the differentiation is performed by a 1-Hz Butterworth filter (RC highpass). Notice the positive-going spikes corresponding to the positive-going NRZ edges and negative-going spikes corresponding to negative-going NRZ waveform edges. In the lower plot, family edge stripping is accomplished with a delay-line of half a bit time and differencer (SUM1), whose output is fed to a Schmitt trigger with a hysteresis of plus-minus 0.5 volt.

The advantage of the bottom plots is that it shows the latter method of edge stripping to be more robust than the former method using simple differentiation although it can be used quite well in a wideband high signal-to-noise (SNR) environment.

DOI: 10.1201/9781003088547-10

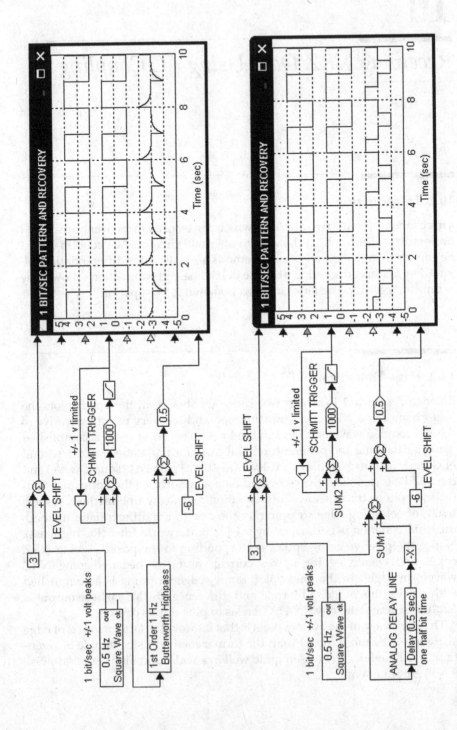

FIGURE 10.1
Two edge stripper examples.

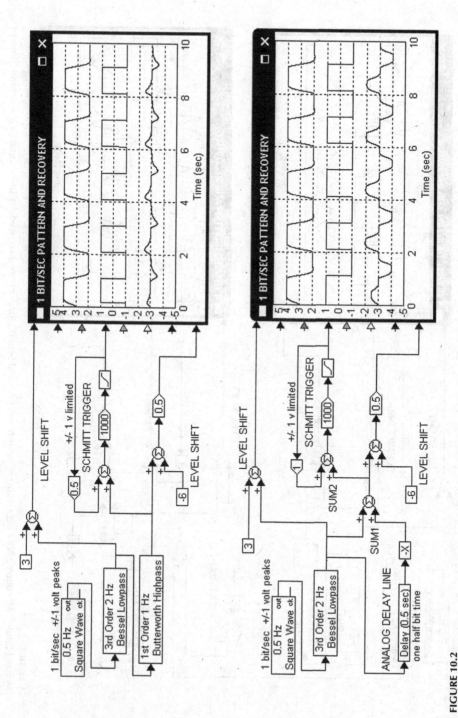

FIGURE 10.2
Band-limited AC coupled NRZ data recovery via differentiator and Schmitt (upper traces) versus delay line subtractor and Schmitt (lower traces).

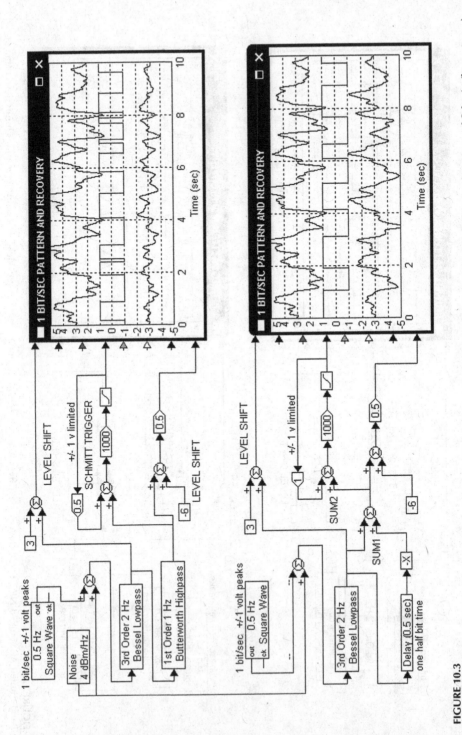

FIGURE 10.3
Noisy band-limited AC coupled NRZ data recovery via differentiator and Schmitt (upper traces) versus delay line subtractor and Schmitt (lower traces).

10.3 Delay Line and Differencer

It is interesting when a 3rd order 2-Hz Bessel lowpass filter is added to represent a band-limited channel effect on the 1 Bit per second data for the example in all plots in Figure 10.2. The differentiation method of the upper plot group clearly has suffered a reduced peak amplitude by a factor of two. But it is still usable after scaling the Schmitt trigger to have half its former hysteresis window. In contrast, the delay line and differencer method is still robust and requires no change in design center value of hysteresis.

10.4 SNR Considerations

Consider Figure 10.3. Additive white noise has been summed with the data to represent receiver front-end noise or signal attenuation. Care was exercised to ensure that the same noise waveform was used in both the case of differentiation and delay line and differencing. Clearly, the delay line case is still more robust and suffers fewer errors than the simple differentiation method. In all cases, one might want to hold the peak-to-peak waveforms constant either through hard limiting or automatic gain control (AGC) so that the Schmitt trigger hysteresis remains appropriate.

11

Gilbert Gain Cell Versus RF Mixer

11.1 Introduction

The balanced modulator or radio frequency (RF) mixer featuring four-quadrant operation, which has been implemented in several integrated circuit versions is described in this chapter. It is generally linear only at one port, the other port being overdriven into switching mode. With the addition of a logarithmic current-to-voltage stage this can be made linear at all ports. The Gilbert gain cell and two- and four-quadrant linear multiplier are described in this chapter, along with the traditional "plain vanilla" Gilbert gain cell.

11.2 Balanced Modulator or RF Mixer

The circuit of Figure 11.1 is a balanced modulator (mixer). It is four-quadrant and is normally driven at the LO port by a square wave causing commutation switching of the upper transistors. The LO port (V3 node 8) is very nonlinear and not suited for purely an analog multiplier application for that reason. However, a mere 300-millivolts peak LO signal will commutate the upper transistors (Q3, Q4, Q5, Q6) from nearly full on to full off. The baseband signal V2 (node 7) is a 1-MHz sinewave applied to differential amplifier transistors Q1and Q2 and is upconverted by the local oscillator (LO) V3 signal of 10 MHz in this example, so that the output at node 10 ("diff") consists of a double-sideband suppressed carrier (DSBSC) result, namely 9 MHz and 11 MHz (Figure 11.2). V1 is a bias voltage equal to the offset voltage of V2 to balance the differential amplifier transistor pair Q1–Q2. If V1 is set to some other direct current (DC) voltage, the 10-MHz LO carrier will not be suppressed. This action might be useful, for example, to generate standard amplitude modulation (AM).

DOI: 10.1201/9781003088547-11

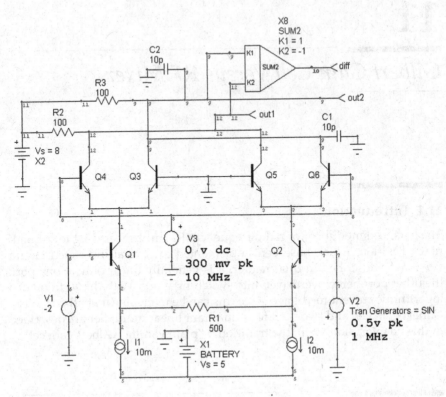

FIGURE 11.1
Schematic of balanced modulator; FC (carrier frequency) represented by V3 = 10MHz; FX (frequency resulting in sidebands) represented by V2 = 1 MHz.

11.3 Gilbert Gain Cell and Linear Multiplier

Before discussing the Gilbert cell [33] and linear multiplier, we need to discuss the output current at each output of a differential pair. Look at Figure 11.3. Two bias currents each having a value of I_y are modulated by voltage Y applied differentially to the two bases of transistors QA and QB to produce collector currents, respectively, of $(I_y + I_yY)$ and $(I_y - I_yY)$. Picture another identical circuit where subscript y is replaced with subscript x and Y is replaced by X (and possibly a PNP version in order to flip the polarity of the collector currents). Figure 11.4 shows the collector currents versus Y for a 1 mA value for current sources I_y and resistor R1 of 1K ohms.

If this differential amplifier output current pair is applied to the multiplier core of Figure 11.5, a linear four-quadrant multiplier results. Diode-connected

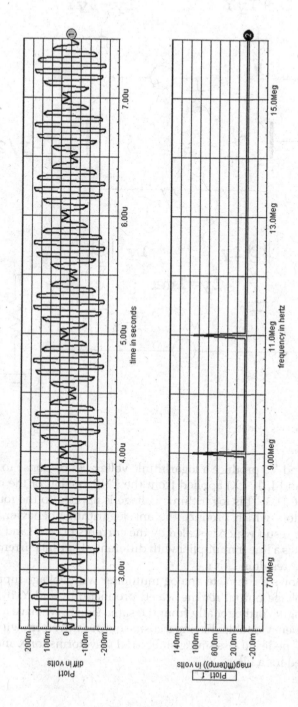

FIGURE 11.2
Balanced modulator fast Fourier transform (FFT); FC = 10 MHz; FX = 1 MHz.

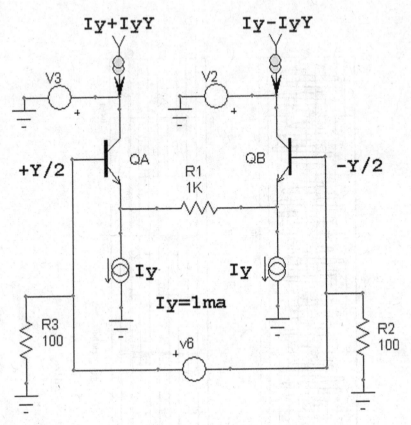

FIGURE 11.3
Diff amp output currents model.

transistors V1 and V2 produce a logarithmic voltage in response to the currents $I_x(1 + X)$ and $I_x(1 - X)$ applied from the PNP version of the diff amp shown in Figure 11.3. This logarithmic voltage is applied to the four cross-coupled transistors which produce the anti-logarithmic behavior, thereby forming a linear result which is scaled by the currents $I_y(1 + Y)$ and $I_y(1 - Y)$.

The end result is a linear multiplier with differential output currents which when subtracted result in 2kXY.

Figure 11.6 shows a complete analog multiplier with voltage inputs for X and Y and a voltage output for the scaled product of X and Y. By driving both X and Y inputs orthogonally from the same signal, we have a squarer, useful as a frequency doubler having a cosine output waveform without any concomitant DC pedestal (assuming sinusoidal waveforms for X and Y) and a minimum need for a filter.

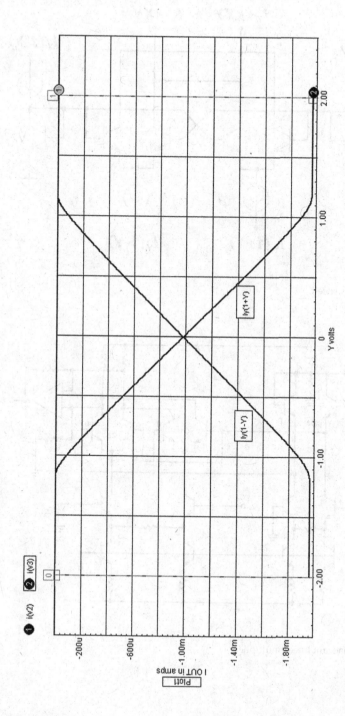

FIGURE 11.4
DC sweep of diff amp.

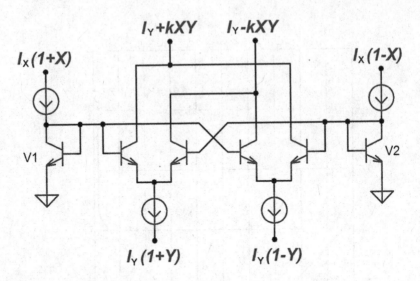

FIGURE 11.5
Linearized Gilbert cell.

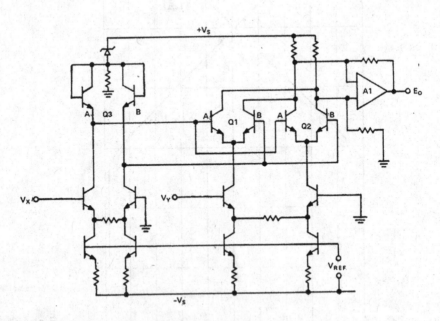

FIGURE 11.6
Actual four-quadrant linear multiplier.

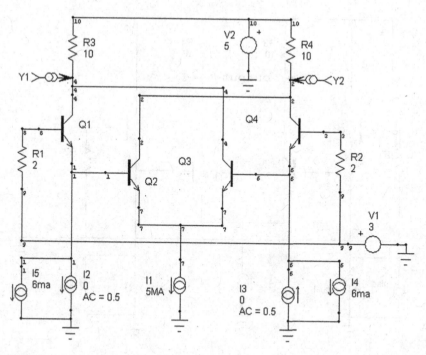

FIGURE 11.7
Plain vanilla Gilbert gain cell schematic.

11.4 "Plain Vanilla" Gilbert Cell

A Gilbert gain cell exists in the form of a current conveyor as in Figure 11.7. Transistors Q1 and Q4 act in a dual role: as common-base stages and also logarithmic current-to-voltage converters. Transistors Q2 and Q3 act as anti-logarithmic stages which sum their output currents with the Q1 and Q4 outputs, thereby not wasting the common-base output signal currents. Differential input signals are I2 and I3 and final output signal currents are Y1 and Y2. The final current gain of the cell is amplified by Q2 and Q3, plus the original input currents. If I1 is set to zero, the output currents Y1 and Y2 are simply equal to I2 and I3, respectively. This circuit can be cascaded with fairly good preservation of gain-bandwidth product.

The circuit's purpose is to yield a current conveyor with a minimum gain of one.

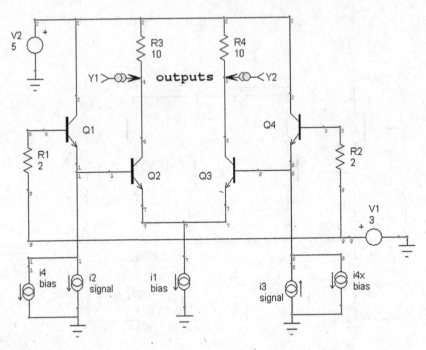

FIGURE 11.8
Variable gain or attenuation gain cell.

If a voltage-controlled amplifier (VCA) is desired that has the ability to be programmed for a gain range that includes attenuation as well as gain, the mere re-connection of the collectors of Q1 and Q4 to a fixed positive rail will accomplish this (see Figure 11.8). The current gain will be programmable by varying i4 as well as i1.

12

Passive Components

12.1 Introduction

This chapter discusses actual physical resistor construction, specifically thick versus thin film, wire-wound and non-inductive, and capacitor types, film and single layer and ceramic, inductors both ordinary and conical, resonators, and the microphonics problem. An equation is given to relate the microphonic sidebands resulting from the gamma of the resonator, typically surface wave or quartz.

12.2 Resistors

Resistors can have undesirable characteristics such as high inductance (e.g., wire-wound resistors) or high capacitance (thick film) or large variation of resistance with temperature (thick film) or accuracy or aging.

Figure 12.1 shows how a non-inductive (or minimally so) wire-wound resistor is made. Two solenoid helix layers are wound in opposite directions, thus the flux from layer 1 cancels the flux from layer two.

Resistors inside an integrated circuit can be made of material which has hugely sloppy tolerances, sometimes hundreds of percent, but the ratio of those resistors to each other can be exquisitely stable and accurate. For this reason, designers of integrated circuits (ICs) are conscious of techniques to exploit this property.

However, several IC foundries offer on-chip resistors made from other materials or metals, which have very reasonable accuracy and stability (for example, some gallium arsenide integrated circuits – GaAs IC – processes).

Discrete resistors can be thin or thick film based. Thick film resistors can be laser trimmed to very accurate values but these will drift horribly over temperature (as bad as 200 parts per million per degree Kelvin). They also exhibit high capacitance (0.3 to 1 pF). Their construction is shown in Figure 12.2.

DOI: 10.1201/9781003088547-12

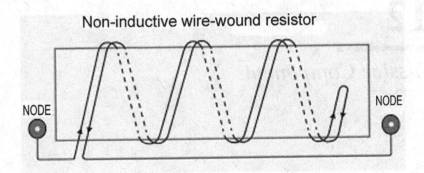

FIGURE 12.1
Non-inductive wire-wound resistor.

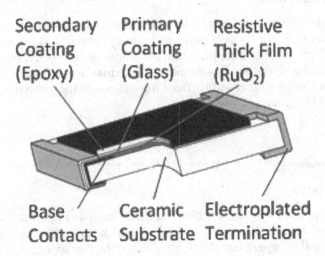

FIGURE 12.2
Thick film surface mount resistor construction.

Thin film resistors (Figure 12.3) are made by metal sputtered onto a substrate and exhibit good stability over temperature and good accuracy, and can be laser trimmed.

12.3 Inductors

Most inductors are intended for surface mount applications. Most (even multi-layered) self-resonate well below microwave frequencies. Often when a bias choke is required, it must have a large inductance value in order to provide large impedance at the lowest frequencies, but unfortunately this makes its parasitic capacitance large, thus becoming challenging at the

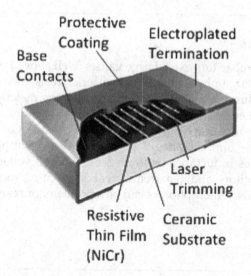

FIGURE 12.3
Low-frequency thin film surface mount resistor construction.

highest frequencies. One solution is to string several inductors in series, thereby distributing their capacitances in series fashion, but this can breed layout-related (and other) parasitic capacitances. A better solution is the conical wire-wound inductor (Figure 12.4).

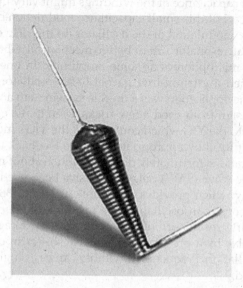

FIGURE 12.4
Conical inductor (Coilcraft).

12.4 Capacitors

Several types of capacitors exist. Large values (well above 1 micro Farad) are electrolytic and large physically. Effective series resistance (ESR) is often a consideration. Aluminum and tantalum are common materials.

At higher frequencies (1 KHz to 10 MHz), film capacitors are common and usually available in surface mount compatible construction. At microwave frequencies, these film caps begin to exhibit disappointing features such as excessive series inductance or limited breakdown voltage. In that case, single-layer capacitors (and in extreme breakdown cases, vacuum capacitors) are often used. Also, microphonics can be a problem for many, especially for ceramic caps.

12.5 Resonators

Often, we need a resonator (aka "tank circuit") in an oscillator application. Lowest cost usually favors an inductor in parallel with a capacitor. However, in addition to poor Q (high dissipation) these tanks will have poor immunity to microphonics (mechanical vibration stimulus converted to electrical response). Often the inductor windings are encased by epoxy to limit the movement of the coils, but this is generally inadequate.

The distributed capacitance of the windings might vary up to 0.1 pF under vibration influence, but this small capacitance modulation can cause a large relative frequency modulation of the oscillator (in the KHz or MHz range).

A quartz crystal resonator offers better mechanical vibration immunity but still exhibits microphonics at some amplitude. In one case, a defense contractor designed a yttrium-iron garnet (YIG) oscillator, which yields a fantastic Q, but the application was a missile system with acute sensitivity to vibration. The design team tried a few tricks, even to the extent of gluing a tiny microphone to the YIG which could sense the YIG's mechanical disturbance and correct for it by appropriate negative feedback techniques. Most pedestrian applications fortunately don't care much about microphonics.

Surface acoustic wave (SAW) delay lines have been used in oscillators to exhibit time delay which provides a multiple of 180- or 360-degrees phase shift for application in an oscillator. The result is an effective Q not quite as good as a quartz resonator but possessing the advantage of having this phase shift variable by a programmable phase shifter in order to shift the frequency of oscillation. Frequency tunability can be as high as +/− 200 ppm using this method.

12.6 Computing Microphonics Due to Sinusoidal Vibration

Most oscillators have some vulnerability to microphonics. At the highest microwave frequencies, those oscillators using the highest Q resonators include those using dielectric resonators, YIG, cavity, and fiberoptic resonators and frequency-multiplied SAW and quartz. Consider a missile which has a built-in frequency reference (e.g., a frequency synthesizer). The synthesizer will most likely employ a quartz crystal reference at 10 MHz or so, multiplied by the synthesizer structure. During take-off, the missile is accelerating and any vibration filter, such as a sling in which the oscillator is suspended, will be slammed against a mechanical stop, rendering that mechanical filter lowpass benefit zero. In the steady state, the sling suspension lowpass filter starts to benefit the quality of the oscillator signal. We can compute relative quality of oscillator signal by obtaining the gamma of the oscillator. Sinusoidal vibration in a particular axis will cause sidebands on our signal which follow the following expression:

$$S = 20\log[(a\Gamma f_o / \{2f_v\})\tag{12.1}$$

where:

S = sideband level in dbc
Γ = acceleration sensitivity, a property of the oscillator in g^{-1}
a = peak magnitude of the vibration force in g's
f_o = final frequency of oscillator (after multiplication if used) in Hz
f_V = vibration frequency in Hz

In the event that Γ is not supplied, one can measure S, then re-write Equation 12.1 to solve for Γ.

A typical value for Γ is 10^{-9} per g for a SAW oscillator at 2 GHz.

13

Unwanted Sidebands Effect on Adjacent Channel(s)

13.1 Introduction

Unwanted sidebands on a local oscillator signal can be discrete frequencies (spurs) or phase noise. In either case, adjacent channels may be dragged into the receiver passband and filtering must be done at radio frequency (RF) front end rather than (intermediate frequency (IF) band) to escape the problem. This is true no matter what modulation format is used.

13.2 Explanation

Whether unwanted sidebands are due to vibration, stray crosstalk, or phase modulation or phase noise, the impact on performance of a receiver is a possible issue, depending on magnitude. Consider an example. A signal of 1 GHz is applied to a receiver with a 70 MHz IF, using a 1.07-GHz local oscillator. Next assume that the LO has a sideband 100 MHz away from L.O. caused by incidental sinusoidal phase modulation. Next assume that this sideband is down at -60 dBc. Also let us assume that at IF we can tolerate an interfering signal no stronger than -25 dBc. Then a 1.1-GHz signal entering our unfiltered receiver front end at 35 dBc above our reference RF level (usually the minimum signal level expected) will jam the center of our desired channel. Filtering at our intermediate frequency stages will not help. This is why we worry about sidebands on a local oscillator. The threat is similar whether or not the sideband energy is sinusoidal. Phase noise is an example. Ample references exist to forecast the phase noise amplitude and spectral shape of an oscillator. [17, 34].

DOI: 10.1201/9781003088547-13

14

Injection Locking

14.1 Introduction

Injection (inj) locking and pulling are phenomena that can be destructive or helpful depending on the application. In particular, when an oscillator is exposed to another oscillator's presence, it may lock onto the other oscillator's frequency. The classical Adler equation is presented to describe the frequency range over which locking can occur depending on the aggressor oscillator's amplitude and offset frequency. A simulation is shown to illustrate. Injection locking and phase locking are often incompatible because the phase angle reached by the phase-locked loop may be different than the angle sought by injection lock.

14.2 Details

The tendency of one frequency source to attempt to slave the frequency of another is well documented. [35–37]. This tendency becomes increasingly powerful as the two sources' initial frequencies converge to some number, call it Δf_O whereupon the master can lock or strongly influence the slave. If the master is also an oscillator physically located nearby, there can exist mutual influence. For example, expect surprise when trying to slave a 100-watt magnetron to a 10-watt solid state oscillator. The 10-watt oscillator will likely become the slave. Thus, an isolator or buffer amplifier of some kind will be required at the output of the weaker oscillator trying to feed the stronger one unless one intends the anomaly (unlikely).

Refer to Figure 14.1 which shows the block diagram and spectrum of an oscillator comprised of a tank circuit of 20% bandwidth (low Q = 5) in the form of a second order Butterworth filter centered just below 100 MHz, a radio frequency (RF) amplifier with 10 dB gain (and internal compression) and a positive feedback summer allowing a fixed signal frequency of exactly 100 MHz to enter the oscillator with attenuation. Figure 14.1 also shows 90

DOI: 10.1201/9781003088547-14

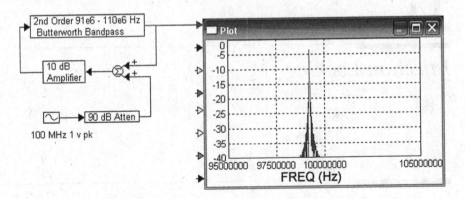

FIGURE 14.1
Free running oscillator not injection locked.

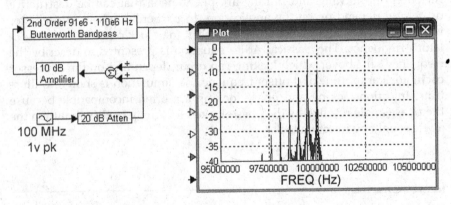

FIGURE 14.2
Injection lock attempt using 20 dB attenuation.

dB for this attenuation, making the 100-MHz signal effectively invisible to the oscillator. Note that the oscillator is free running at a frequency just under 100 MHz, but in Figure 14.2, the attenuation has been reduced to 20 dB, causing the oscillator to attempt to slave, but only pull closer to 100 MHz, accompanied by spectral splatter (extra signal frequencies). Finally, in Figure 14.3, we see that the attenuator has been set to only 10 dB, injecting more amplitude from the 100-MHz source, causing the oscillator to lock enthusiastically to 100 MHz. So much for qualitative behavior. Robert Adler derived quantitative behavior equations describing the important relationships between the amplitude of the injection signal, its frequency displacement from the oscillator's free running frequency, and the Q of the oscillator tank circuit.

$$\Delta f_O = \left(\frac{f_O}{2Q}\right)\left(\frac{A_{INJ}}{A}\right) \tag{14.1}$$

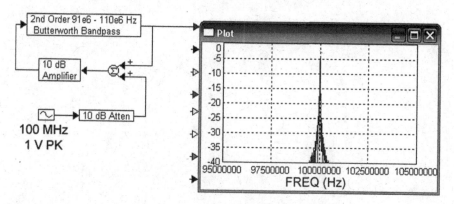

FIGURE 14.3
Injection lock success using 10 dB attenuation.

where

Δf_O = the difference between the oscillator frequency and the injection frequency for locking to take place.

f_O = the resting frequency of the oscillator in Hz prior to lock

$\dfrac{A_{INJ}}{A}$ = the ratio of injection voltage to normal oscillator amplitude.

In our case for Figure 14.3, we plug in the numbers and get the result that for the component values shown, we can assure lock over a range of Δf_O = 3 MHz.

According to this equation, injection locking can occur with progressively increasing ease when either the frequency difference becomes smaller, or the Q becomes smaller, or the injection signal becomes larger. This has nasty consequences if the two signal frequencies are very near each other and locking is not desired. Then we need only a weak injection signal amplitude to slave the oscillator.

One application of injection locking is to make low noise frequency dividers since the slave can be an oscillator of 1/N times the master injection frequency. Another application is to transfer the signal purity of the master to the slave (e.g., clean up the phase noise sidebands of a magnetron to more closely match the lower phase noise of the master signal).

Injection locking or pulling is hard to manage when undesired. In a phase-locked loop (PLL), the master signal may be attempting to injection pull the slave while the PLL is attempting to do likewise. The result is a tug of war because the phase angle to which the PLL desires to settle is not equal to the phase angle satisfying the injection locking, usually because the layout and/or circuit topology has crosstalk that cannot merge the two phases as master frequency is varied. The original paper by Adler [35] is a good place to start to further understanding, as is the excellent paper by Razavi [37].

15

Phase-Locked Loops

15.1 Introduction

Equations relevant to second order type two phase-locked loops (PLL) are presented in this chapter along with a BASIC program for automating the computations. (Second order loops can be devolved to first order, with details left to the reader.) A false lock prevention method is presented which exploits the property of an analog phase detector and a direct current (DC) bias provided by a self-sweep waveform to ignore sidebands or spurs, assuming the loop bandwidth is small enough not to bracket the spur and the main signal when in true lock, in which case no false lock could have occurred anyway. The approach uses a Schmitt trigger to generate a switching voltage which drives a resistor for conversion to a sweep current driving the loop integrator. The sweep current must be smaller than the peak error current from the phase detector when the loop is in true lock.

15.2 The Most Popular Second Order Type 2 PLL

Figure 15.1 shows the classical block diagram of a feedback control system plus equations [38].

The classical control system equation convention is used, with the input compared to the feedback signal via a differencer. In Figure 15.2 this differencer is the phase detector for the PLL and has a conversion gain of K_D volts per radian. The output of the phase detector feeds a loop filter with a transfer function of F(s). The output of said loop filter then feeds a voltage-controlled oscillator (VCO) with a transfer function of K_V/s, where K_V is expressed in radians per second per volt. The phase detector, filter, and VCO together comprise the control system forward gain of G(s). The term H(s) is usually an attenuation factor represented by the term $1/N$ where N = frequency divider factor.

DOI: 10.1201/9781003088547-15

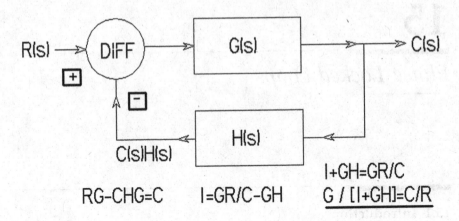

$$1+GH=GR/C$$

$$RG-CHG=C \qquad 1=GR/C-GH \qquad \underline{G / [1+GH]=C/R}$$

FIGURE 15.1
Block diagram of a control system.

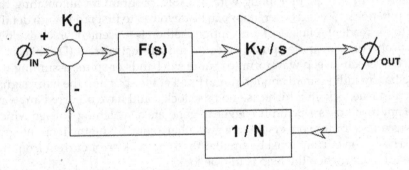

FIGURE 15.2
PLL model block diagram.

Since we are trying to analyze phase and derive the transfer function, we need to know that frequency is the derivative of phase. This explains the form of the VCO transfer function as an integrator for phase, hence $\{K_V/s\}$ is the correct form mathematically.

The equations describing the PLL of Figure 15.2 are:

$$G(s) = [K_D K_V F(s)] / s \tag{15.1}$$

$$\varnothing_{OUT} / \varnothing_{IN} = [K_D K_V F(s)] / [s + K_D K_V F(s) / N] \tag{15.2}$$

where:

$$F(s) = \frac{1 + sR_F C}{sR_1 C} \tag{15.3}$$

for the filter of Figure 15.3

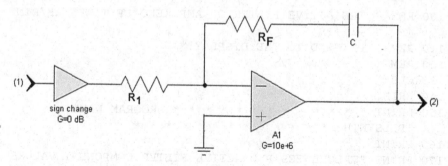

FIGURE 15.3
PLL type 2 loop filter model.

and the loop natural frequency in "radians per seconds" is:

$$\omega_N = \sqrt{K_D K_V / [R_1 C]} \tag{15.4}$$

And the damping (no units)

$$\partial = \omega_N R_F C / 2 \tag{15.5}$$

Notice the square root operator in Equation 15.4. If we err by 20% low or high in our specification of any of the terms, the error in the loop natural frequency will be only 10%. (the square root of 1.2 is 1.095).

According to Egan [39], the optimum damping value (for integrator plus lead) for best settling is 1.

Here is a simple program written in BASIC to automate the calculations (it might be required to tweak the code a little if latest versions of BASIC are an obstacle):

```
10 REM      PROGRAM = PLLSYNTH
20 REM
30 REM      THIS PROGRAM COMPUTES THE REQUIRED COMPONENT
            VALUES
40 REM      FOR RF, RIN, AND CF OF THE ACTIVE LOOP
            FILTER, GIVEN
50 REM      FO (LOOP NATURAL FREQUENCY IN HERTZ), DAMPING
            FACTOR
60 REM      "D", VCO TUNING SENSITIVITY (IN HZ/VOLT) AND TWO
70 REM      DIVISION RATIOS X1 AND X2 (REPRESENTING THE
            TWO EXTREMES
80 REM      OF THE PROGRAMMABLE DIVIDER SETTING, N). IN
            ADDITION,
90 REM      THE FM NOISE FLOOR OUT-OF-BAND DUE TO THE
            NOISE LEVEL OF
```

```
100 REM    THE ACTIVE FILTER OP-AMP (ENOISE TIMES RF/RIN
           +1) IS
110 REM    PRINTED TO THE DISPLAY.
120 REM
130 CLS
140 PRINT
150 PRINT " * * * * * * * * * * * PROGRAM NAME =
    'PLLSYNTH' * * * * * * * * * * * *
160 PRINT
170 PRINT "CALCULATES PLL ACTIVE FILTER COMPONENT VALUES
    AND OUT-OF-BAND FM NOISE"
180 PRINT:PRINT
190 INPUT "VCO SENSITIVITY IN HERTZ PER VOLT    =";V
200 PRINT
210 INPUT "PHASE DETECTOR GAIN IN VOLTS PER RADIAN =";P1
220 V=V*P1
230 PRINT
240 INPUT "REQUIRED LOOP DAMPING FACTOR   D   =";D
250 PRINT
260 INPUT "REQUIRED LOOP NATURAL FREQUENCY  F0  =";F0
270 PRINT
280 INPUT "SMALLEST FREQUENCY-DIVISION RATIO  N  =";X1
290 PRINT
300 INPUT "LARGEST FREQUENCY-DIVISION RATIO   N  =";X2
310 PRINT
320 INPUT "OP-AMP INPUT VOLTAGE NOISE PER ROOT-HZ =";EN
340 PRINT "*********************************************
    ********************************"
360 PRINT "FOR THE CASE WHERE THE DIVISION RATIO N =";
370 PRINT X1
380 PRINT
390 TWOPI=2*3.14159
400 CF=V/(TWOPI*F0*F0*X1*1500!)
410 REM   RIN=1500 OHMS  (FOR LOW JOHNSON NOISE WITH
    MOST OP-AMPS)
420 RF=2*D/(TWOPI*F0*CF)
430 V=V/P1
440 FM=V*EN*(1+RF/1500!):'  IN HZ FM DEVIATION RMS PER
    ROOT-HZ BW
450 PRINT "   CF = ";:PRINT CF;:PRINT" FARADS"
460 PRINT
470 PRINT "   RIN =";:PRINT " 1.5 K OHMS"
480 PRINT
490 PRINT "   RF = ";:PRINT RF;:PRINT" OHMS"
```

```
500 PRINT
510 PRINT "   FM = ";:PRINT FM;:PRINT "HZ RMS DEV PER
    ROOT-HZ FM DET BW (EXCLUDES RESISTOR NOISE)"
511 PRINT "    WELL ABOVE THE LOOP NATURAL FREQ F0
    =";:PRINT F0
520 PRINT
530 PRINT "PULL-IN RANGE FOR NO CYCLE-SLIPPING IS APPROX
    =";:PRINT X1*F0;:PRINT "HZ"
540 PRINT
    "-----------------------------------------------------
    ----------------------------"
560 PRINT "FOR THE CASE WHERE THE DIVISION RATIO N =";
570 PRINT X2
580 PRINT
590 PRINT "SIMPLY CHANGE RIN (OR AN EQUIV EFFECT)
    TO";:PRINT X1*1500!/X2;
600 PRINT " OHMS"
605 PRINT
606 FM=V*EN*(1+RF/(X1*1500!/X2))
610 PRINT "OUT-OF-BAND NOISE FLOOR WILL NOW RISE
    TO";:PRINT FM;:PRINT "HZ DEV RMS PER ROOT-HZ";
615 PRINT "FM DET BW EXCLUSIVE OF RESISTOR NOISE (WELL
    ABOVE F0)"
620 PRINT
630 PRINT "PULL-IN RANGE FOR NO CYCLE-SLIPPING IS APPROX
    =";:PRINT X2*F0;
640 PRINT "HZ"
645 INPUT "TO END, HIT ENTER";A$
650 END:STOP:END
```

The value assumed for R1 in the above program is 1500 Ohms, but low noise op-amps may benefit from much lower values to minimize Johnson noise. In that case, scale all other resistors in the loop filter by the same factor and the capacitor by the inverse of that.

Sidebands on the VCO output may benefit from additional high order filtering [17].

In an analog PLL wherein the phase detector is analog (e.g., non-edge-triggered) the phase noise performance can outperform those PLLs using edge-triggered phase/frequency detectors. The presence of edge triggering aliases the very broad spectrum of self-noise and folds the entire spectrum into the PLL sampling band. The process is described in reference [17] by Daniel Talbot.

There are occasions where an analog approach would be impractical, namely for extreme low bandwidth designs, such as a PLL with 1 Hz f_n.

15.3 False Locking Prevention for Sweeping PLL

An analog phase detector has a conversion gain readily obtained by its operation as a mixer. After all, by operating the mixer with a dominant (LO) waveform at frequency "f1" and feeding the other mixer port with frequency "f2" where f1 and f2 are close together in frequency, we see a beat waveform. The beat waveform is simply a slow phase-walk and reveals the shape and gain of the phase detector. One can measure the slope of the beat waveform at the zero crossings in volts per radian and this number is the phase detector conversion gain.

In Figure 15.4, we have structured a model where two signals are fed to a PLL phase detector (X1), one signal we will call the "exalted signal" (X2) and the other the "spur" (X3) where the amplitude ratio of the two is 4:1. The exalted signal feeds summer X4 at K1 input, set to a gain weight of one. The spur feeds summer X4 at K2 input having a weighting of 0.25. Signal source X5 is the VCO output of the PLL (remaining diagram not shown). For simulation purposes, we filter the phase detector output using a lowpass filter (X6) with a bandwidth wide enough to pass the beat but reject other products of mixing that we are not interested in. If we set X5 to a frequency near that of X3, we see a beat waveform shown in Figure 15.5 (smaller amplitude sinewave). If we set X5 to a frequency close to that of X2 (exalted signal) we get a four times larger amplitude signal, and this is the signal we want to phaselock. A simple false lock prevention measure is to add a DC offset to the phase detector output that is larger than the peak amplitude of the smaller beat waveform caused by the spur. Thereby, the PLL cannot lock to the spur because the spur's zero crossings have now been shifted by the amount of

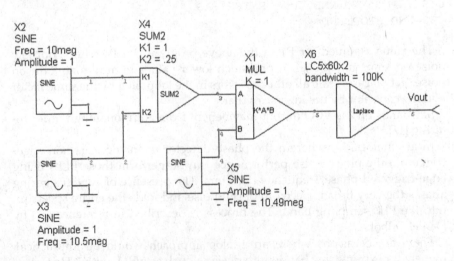

FIGURE 15.4
PLL false lock schematic via DC bias.

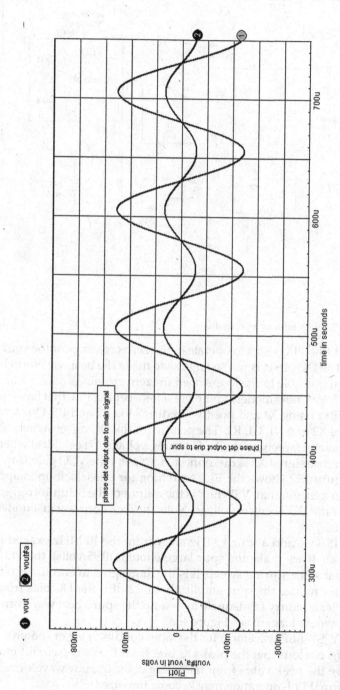

FIGURE 15.5
PLL false lock elimination via DC bias.

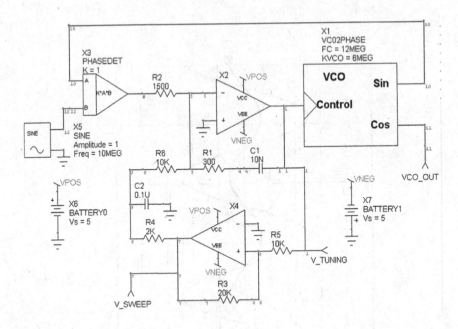

FIGURE 15.6
Self-sweeping PLL with false lock prevention.

the DC offset (the PLL seeks to operate at the zero crossings of the spur beat waveform). The DC offset is not so large as to make the beat waveform of the exalted signal invisible to the loop's need for zero crossings.

In Figure 15.6 we next model a complete self-sweeping PLL that has Schmitt trigger circuitry using X4 and positive feedback via R3 and R5. Our loop filter consists of X2 and R1, C1, R2. The polarity of the DC offset switches back and forth (thereby sweeping the VCO tuning voltage to seek lock) when the 10 MHz reference signal X5 is disconnected. (Note: t the VCO gain is in rad/sec/volt). Figure 15.7 shows the PLL searching for phaselock upon application of the reference signal. VSWEEP is the output of the Schmitt trigger and VTUNING is the VCO control voltage. Note: the lock occurs at just under 600 microseconds.

In Figure 15.8 we add a spur 1 MHz away from the 10-MHz exalted reference signal X5. If we make the spur large enough (0.95 volts), the PLL will false lock on it since, during sweep, it is the first signal to encounter (Figure 15.9). But if we reduce the spur amplitude to -12 dBc, the DC bias from the Schmitt trigger circuitry is adequate to swamp the spur's beat waveform and false lock is avoided (again see Figure 15.9).

Note: the V_SWEEP waveform for the latter condition is not shown

It should be obvious that the peak DC bias from the sweep circuit must be set just above the peak value from the phase detector spur waveform. If set too high, normal PLL operation may become impaired.

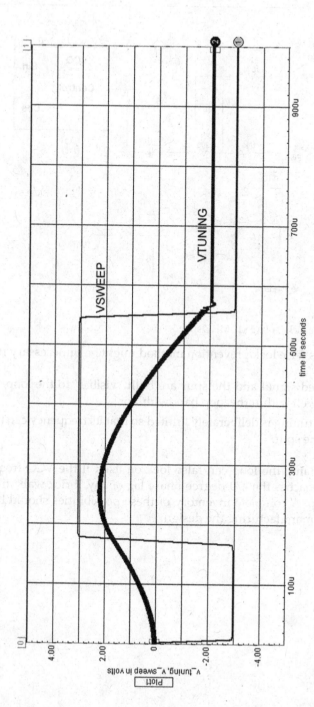

FIGURE 15.7
PLL sweep and lock for VCO error of 2 MHZ.

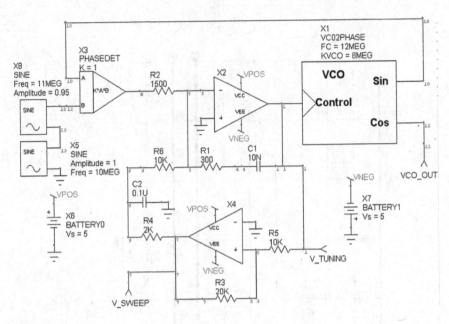

FIGURE 15.8
PLL expecting false lock for large 11-MHZ spur.

Note that this false lock prevention method might be unnecessary if:

(a) The exalted signal and the spur are both "visible" to the loop, i.e., both are well within the loop bandwidth or

(b) The VCO tuning is deliberately limited so that its frequency can't get close to the spur.

On the other hand, the loop can false lock on itself if the VCO frequency excursion approaches the 270-degree phase lag of any "brick wall" filtering within the loop. A complete inventory of these possibilities should be performed before manufacturing the design.

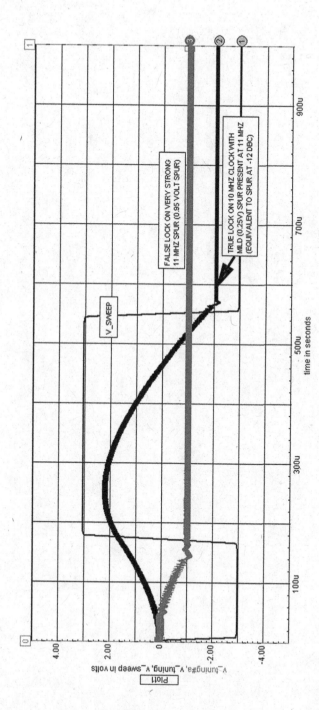

FIGURE 15.9
PLL false locks on large 11-MHZ spur.

16

Distortion Fundamentals and Spectral Regrowth

16.1 Introduction

In this chapter, second and third order distortion ramifications of two-tone stimuli are presented, as is the description of intermodulation, cross-modulation, and spectral regrowth phenomena. The "one deciBel compression point" is explained and used to replace the concept of third order intercept point.

16.2 Second Order Distortion

Let the amplifier, mixer, analog-to-digital converter or other device processing our signals have some weak second order distortion. Mathematically, the gain of the device deviates from a straight line for E_{out} over E_{in}.

Let the output be expressed as follows:

$$E_{OUT} = E_{IN}K_1 + K_2 E_{IN}^2 \tag{16.1}$$

If the input consists of just one signal, we can write Equation 16.1 as:

$$E_{OUT} = K_1 \cos(\omega_1 t) + K_2\{0.5\cos(2\omega_1 t) + 0.5\} \tag{16.2}$$

Notice that because of a mild second order bend from a straight line we generate the original signal scaled in amplitude by K_1 plus a second harmonic plus a direct current (DC) voltage exactly equal to the amplitude of the second harmonic. There may be alternate current (AC) coupling after this output but the DC shift definitely exists prior to the AC coupling.

DOI: 10.1201/9781003088547-16

Next, suppose we have not one but two different signals with differing frequencies to be processed by the same Equation 16.1. Let the first and second signals be expressed as:

$$E_{IN} = A\cos(\omega_1 t) + B\cos(\omega_2 t) \tag{16.3}$$

Then

$$E_{OUT} = K_1\{A\cos(\omega_1 t) + B\cos(\omega_2 t)\} + K_2\{A\cos(\omega_1 t) + B\cos(\omega_2 t)\}^2 \tag{16.4}$$

The resulting products are located in frequency at:

$$(\omega_1 + \omega_2), (\omega_1 - \omega_2), (2\omega_1), (2\omega_2) \tag{16.5}$$

The classical sum and differences frequencies are the result of second order distortion while processing two signals of differing frequency. If we apply two signals at 10 KHz and 11 KHz, we will reap new signals at 1 KHz and 21 KHz (sum and difference) plus the second harmonic of each of the two original signals.

16.3 Third Order Distortion

In the case of a symmetrical transfer function, we obtain third order products plus the modified fundamentals.

If K3 is positive, the transfer function gain increases as we drive the system with signals of increasing amplitude, and if K3 is negative, we have compression as signal amplitude increases. The latter case is the most common for devices like amplifiers and mixers. Ignoring the second distortion terms and fundamentals, using the same two input signals as before, the third order distortion products are:

$$E_{OUT}|_3 = K_3\{A\cos(\omega_1 t) + B\cos(\omega_2 t)\}^3 \tag{16.6}$$

We will leave the result of simplifying the terms to the reader, but the punch line is that if K_3 is negative, the fundamental signals will be compressed by

$$3B^2\frac{K_3}{4} + 3A^2\frac{K_3}{2} \tag{16.7}$$

for the signal at ω_2
  and by

$$3A^2\frac{K_3}{4} + 3B^2\frac{K_3}{2} \tag{16.8}$$

for the signal at ω_1

By inspection, the amplitude of signal A seems to be transferred to signal B and vice versa. This process is called "cross-modulation distortion," not to be confused with "intermodulation distortion" which produces new frequencies as beats. If signal A is amplitude modulated, the *modulation envelope* will transfer to signal B (but upside down).

There will also be third harmonics of each signal, which are generally harmless, plus new cross-products resulting in beats like these:

$$(2\omega_1 - \omega_2), (2\omega_2 - \omega_1), (2\omega_1 + \omega_2), (2\omega_2 + \omega_1) \qquad (16.9)$$

These nasty products are close to being in band. Consider Figure 16.1. The two fundamental frequency signals are at 100 MHz and 105 MHz. They have each been compressed resulting in new sidebands at 200 − 105 = 95 MHz and 210 − 100 = 110 MHz.

From dissecting the third order behavior equations (details in reference [41]), we find that if the fundamental signals are each compressed by 1 dB, the third-order beats depicted in Figure 16.1 will be down at −27.8 dBc.

If we drop both fundamental signals by 10 dB, these beats should decrease by three times this or 30 db. But since the fundamentals have dropped by 10 dB, the relative distortion improvement is only 20 dB.

This is a relationship that is important to remember. Third-order inter-modulation sidebands drop 2 dB for each 1 dB-drop in fundamental signals. Contrast this with 1dBc improvement for each 1db reduction of signals when the second order behavior is dominant (and beats will be simple sum and difference frequencies).

Notice from Figure 16.1 that the third order sidebands are somewhat obedient to our math predicting −27.8 dBc sideband levels when fundamentals are in the neighborhood of 1 dB compressed. Reducing the fundamentals by 20 dB will result in −67.8 dBc sidebands. We have deliberately ignored the concept of third-order intercept point that is described in other texts.

An interesting effect results in digital communication systems where the pseudo noise radio frequency (RF) signal is not linearly processed. Take the case of Figure 16.2 in which a lowpass filtered noise signal is applied to a hard limiter (example of a cubic or odd-order distortion). Since noise can be thought of as an infinite number of sinusoids, there is ample opportunity for third-order intermods between noise frequency A and noise frequency B to create noise-cross-noise cross-products.

In Figure 16.3 we see a more common situation in which a digital communications signal (e.g., 64 quadrature amplitude modulation – QAM) is bandpass filtered and then fed to a power amplifier for final transmission. Notice that the original spectrum after encountering the non-linear amplifier is altered by noise-cross-noise products that cause the spectrum to spill over its assigned frequency bounds (and into an adjacent channel). The model we show is not totally accurate because real Gaussian noise consists of all

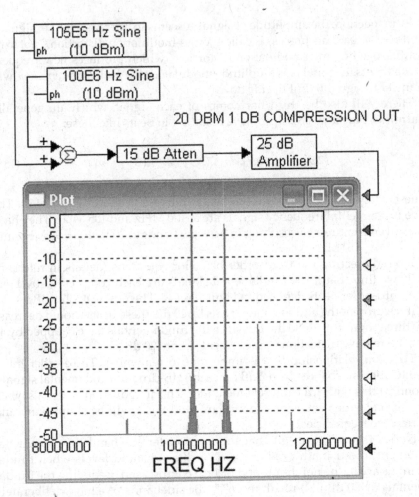

FIGURE 16.1
Two-tone spectrum compression 1 dB.

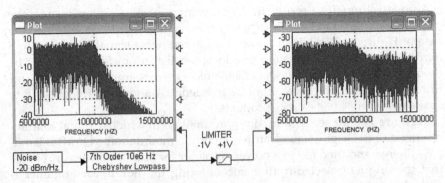

FIGURE 16.2
Example of spectral regrowth.

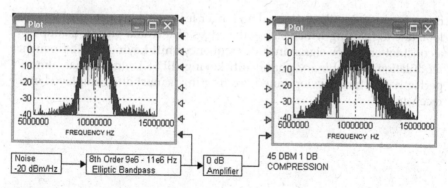

FIGURE 16.3
Example of spectral regrowth in bandpass case.

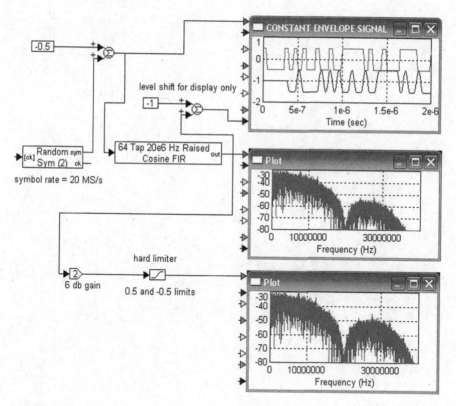

FIGURE 16.4
Example of hard limited lowpass case for bi-phase data.

voltage levels whereas digitally transmitted data does not (unless smeared by too much filtering).

Figure 16.4 shows what minimal harm occurs to a constant-envelope signal as it passes through a nonlinear amplifier. The unwanted spectral products occur because the filtered signal no longer features a perfect constant envelope due to filtering which slopes the edges of the data pulses. Examples of a constant-envelope signal include frequency-shift keying (FSK), frequency modulation (FM), binary phase shift keying (BPSK), etc. FM has almost a perfectly constant envelope and hard limiting is most often employed in the receiver.

17

Optimization

17.1 Introduction

An overview of optimization techniques is presented in this chapter. First, the reader is walked through a systematic optimization of the group delay flatness of a two-stage and a three-stage delay-equalized Cauer bandpass filter (BPF). Second, the reader is shown how to simultaneously optimize several component values at once for the passband transmission factor, passband rolloff, and stopband depth for a single-terminated lowpass filter (LPF).

17.2 Introduction to Curve Flattening

Powerful computer-aided design (CAD)/computer-aided engineering (CAE) software exists that can automatically search and vary component design values to get the best curve fit to flatten a curve or minimize insertion loss or optimize some other property of, for example, a filter. We will choose passive filters for our examples but with the right software active circuitry can also be addressed. In the first discussion we will employ both passive and active circuitry to illustrate how to optimize the group delay flatness of a filter joined with allpass section(s). Passive filters will be used when possible because they don't oscillate and they have very low sensitivity to component variation (because they are essentially ladder networks).

We start with a fifth order Cauer bandpass filter shown at the top of Figure 17.1. It is centered at 1 MHz, and its group delay and transmission magnitude are shown in Figure 17.2. We would like to greatly improve the group delay flatness and have therefore added circuitry to Figure 17.1 consisting of allpass stages using operational amplifier stages A2 and A3. These are both second order delay equalizers. Amplifier A1 is a simple voltage follower to buffer the next stage. Note that port 1 is a special source termination of 50 Ohms and thus we cannot observe voltage, but rather S11 (not plotted) and S21 in measuring delay or transmission. Group delay is displayed on the right vertical axis of Figure 17.2 as GD[S31] in nanoseconds. Transmission is displayed on the left vertical axis as DB[S21]. Notice that group delay is not very flat in the filter's passband, being very saddle-shaped at the center of the

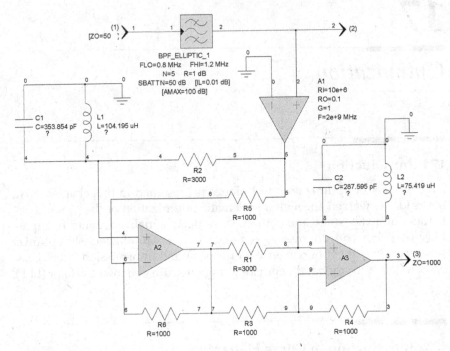

FIGURE 17.1
Schematic of Cauer BPF and group delay equalization circuit (EQ CKT).

band. This is the issue at hand which we shall improve via component value optimization in the equalizer.

First, we need to tell the software which component values it is allowed to vary in order to execute the optimization. For this particular software we simply mark said components with a question mark immediately preceding each value shown in Figure 17.1. These comprise L1, L2, C1, and C2. The question mark is enabled by checking a box in software for each starting value allowing tuning. We can also limit the range of values allowed. It is wise to force the tank circuits to resonate near the center of the filter passband before engaging the optimization routine so that the optimizer can begin near a reasonable vicinity. That would suggest initial values for L1 and L2 of 100 uH and C1 and C2 of 250 pF. The final values after optimization are shown on the same schematic.

Figure 17.3 shows the data entry screen for the optimizer. The first column asks the user to name what property (measurement) will be the subject of optimization. We entered "GD[S31]." The next column is optional (we entered DELAY). The next column asks for a special character and we chose "flatness" represented by a % symbol. The next column is a target value, and we chose a very strict value of 0.1.

For "weight" we chose a value of 1, which in the absence of any other parameters to optimize is as strong as possible. Next, we enter fmin and

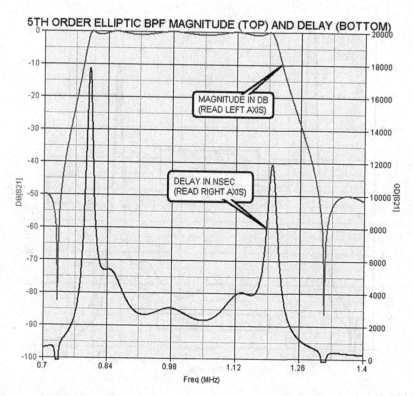

FIGURE 17.2
Magnitude response and delay of fifth order Cauer BPF.

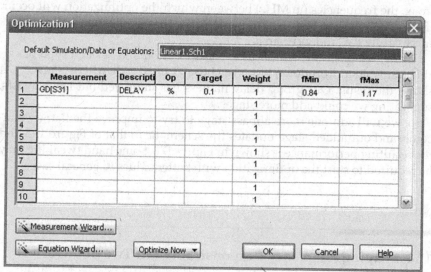

FIGURE 17.3
Optimization screen of Cauer BPF and group delay EQ CKT.

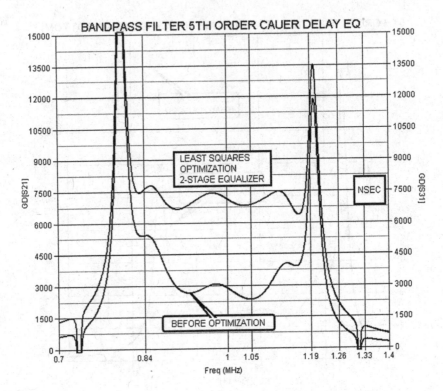

FIGURE 17.4
Two-stage optimized delay EQ for fifth order Cauer BPF.

fmax, the frequencies (in MHz) between which the optimization will occur. If we widen the gap between fmin and fmax, the equalization will not be as successfully flat, and vice versa.

Finally, we push the "OPTIMIZE NOW" button and select a type of calculation from a drop-down box. We chose "LEAST SQUARES." Figure 17.4 shows the result. Our group delay flatness has improved but ripple exists, suggesting we could add more stages.

So, we will add one more allpass stage to try to improve the flatness outcome, and our added stage updates the schematic to that of Figure 17.5 and the resulting performance is shown in Figure 17.6. Some hand tweaking may be required to start the optimization with better values in general cases.

17.3 Shaping Frequency Response between Two Boundaries

Observe the frequency response of a single-terminated lowpass filter in Figure 17.7 prior to optimization. Its schematic showing pre-optimized

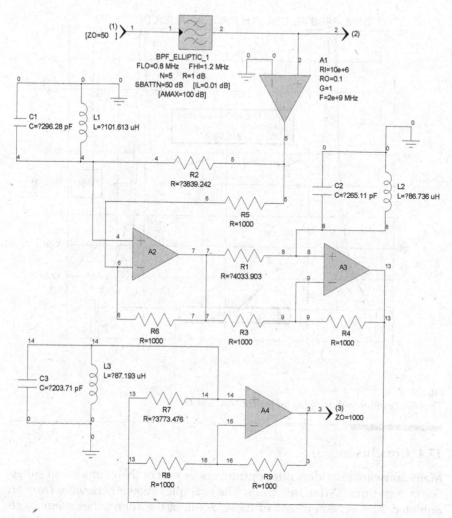

FIGURE 17.5

Schematic of three-stage optimized delay EQ for fifth order Cauer BPF.

component values is given in Figure 17.8. We will next enter our desired optimization targets as in Figure 17.9, then run minimax calculations from the optimizer drop-down button.

The result appears in Figure 17.10 with new component values shown in Figure 17.11 next to each question marked variable-permitted part.

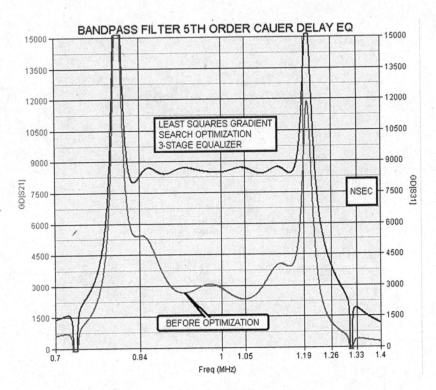

FIGURE 17.6
Three-stage optimized delay EQ for fifth order Cauer BPF.

17.4 Conclusion

Many software providers offer optimizers as part of their simulation suites. Some even offer 30-day free trials. The examples shown here were from an outdated 2004 version of one of those. Some of the approaches seem "seat-of-the-pants" because in fact they are. But an experienced engineer with lots of careful observation abilities can interpret and intervene in the converging process of simulation and optimization. For example, if the designer sees that a component value is heading in the direction of extremely large or small, it may be that that section of the network is unnecessary or that the topology is in serious need of review, or that limits on component value ranges need to be set in the schematic.

There are so many more topics and topologies of filters, networks, and circuits that they cannot all be covered in this book, but it is hoped that the reader has a good starting point from the information presented here. Most important, it is hoped the reader has found the information practical and immediately useful.

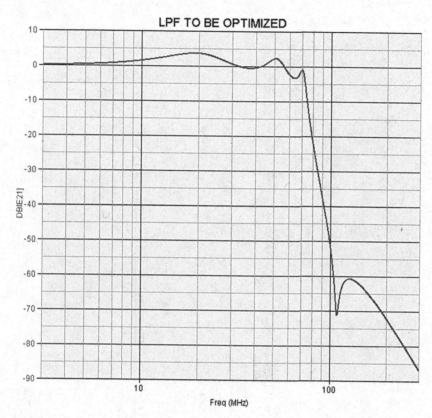

FIGURE 17.7
Single-terminated LPF to be optimized.

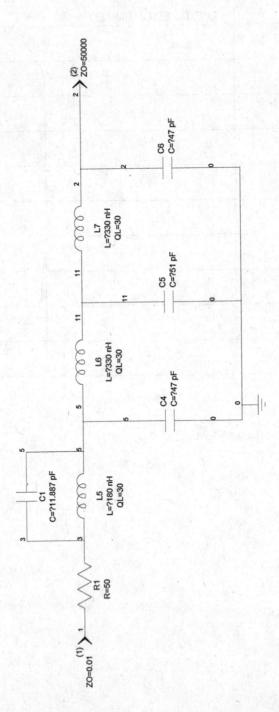

FIGURE 17.8
Pre-optimized schematic of single-terminated LPF.

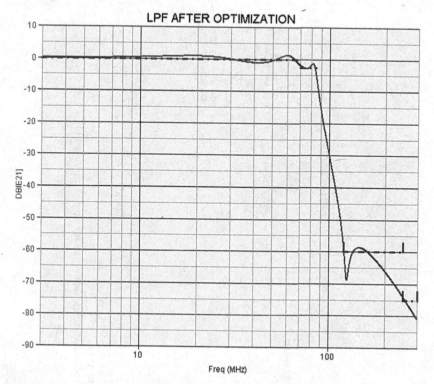

FIGURE 17.9
Single-terminated LPF optimization targets.

FIGURE 17.10
Single-terminated LPF after optimization.

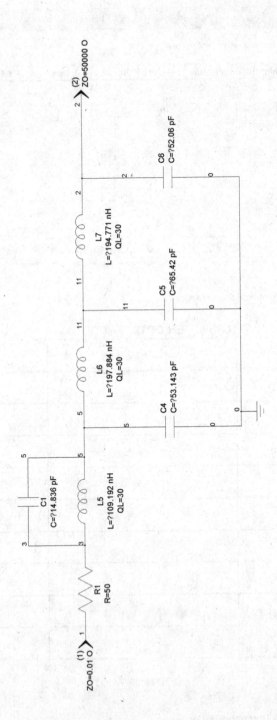

FIGURE 17.11
Post- optimized component values for single-terminated LPF.

18

Quadrature Distortion and Cross-Rail Interference

18.1 Introduction

In Section 8.16, we showed one method of impedance matching a filter using inductors as a means toward improving the frequency response symmetry of a filter. We shall explain in this chapter the reasons to want symmetry. We begin by demonstrating how a passive filter's asymmetry determines non-linear distortion of a sine wave modulation envelope upon envelope demodulation and how to preserve the modulation waveform integrity by using synchronous detection. Subsequently, we show that the same effort to assure symmetry is required to minimize orthogonal leakage in in-phase and quadrature (I/Q) demodulation (e.g., quadrature amplitude modulation – QAM) also called cross-rail-interference.

18.2 Standard Amplitude Modulation (AM) Broadcast Reception with Sideband Asymmetry

Our first example is standard amplitude modulation broadcast transmission and reception.

The model for standard AM is shown in Figure 18.1. Note that two sidebands are generated by the modulation, shown as counter-rotating vectors in Figure 18.2. These sidebands are superimposed on the carrier vector which rotates at ω_c while the modulation sidebands are two, each counter-rotating and superimposed on the carrier vector; the lower sideband rotates at $\omega_c - \omega_m$ and the upper sideband rotating at $\omega_c + \omega_m$ The equation for standard AM with such sinusoidal modulation is:

$$\{\cos(\omega_c t)\}\left[1 + m\cos(\omega_m t)\right] \tag{18.1}$$

DOI: 10.1201/9781003088547-18

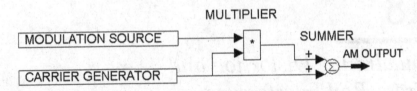

FIGURE 18.1
Standard amplitude modulation model.

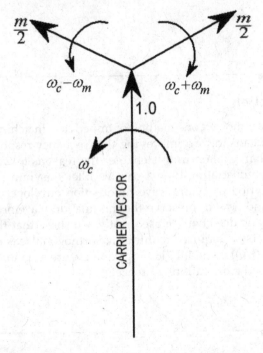

FIGURE 18.2
Standard amplitude modulation with symmetrical sidebands.

which can be written as:

$$\{\cos(\omega_c t)\} + \frac{m}{2}\cos(\omega_c - \omega_m)t + \frac{m}{2}\cos(\omega_c + \omega_m)t \qquad (18.2)$$

where

    m = modulation index (value of between 0 and 1)
    ω_c = carrier frequency, radians per second
    ω_m = modulation frequency, radians per second
    If we "de-spin" the carrier vector and its sidebands, we have a vectorial picture of the total in Figure 18.3.

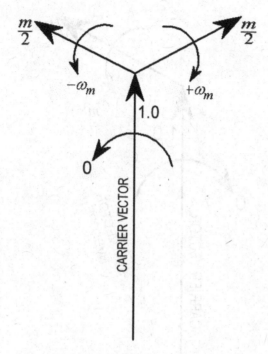

FIGURE 18.3
Standard amplitude modulation with symmetrical sidebands de-spin.

So far, we have not processed the AM signal through a filter. Suppose we add one and the filter is so extremely asymmetrical that the lower sideband is completely attenuated. We then have single sideband non-suppressed carrier (SSBNSC) represented in vector form by Figure 18.4 and the resultant sum vector now exhibits phase modulation (PM) as well as AM (there is now a projection of the sum vector onto both the "Q" (quadrature phase) axis as well as the "I" (in-phase axis). Since the I axis contained the actual original sinusoidal modulation before filtering, its recovery would be the intended modulation, but the sum vector is no longer sinusoidal. Hence, envelope demodulation recovers a distorted version due to filter asymmetry, although admittedly we have chosen to demonstrate the exaggerated case. This distortion is called quadrature distortion and is the pitfall of envelope detection. Notice that synchronous detection does not exhibit this distortion, since it demodulates along one axis only. But synchronous detection requires a separate recovery of the carrier. Such recovery may imply a narrow passband filter or tank or phase-locked loops (PLL) and hard limiter. A suitable model of a synchronous detector is shown in Figure 18.5. If the bandwidth of the recovery circuit is too wide, the demodulation will devolve into envelope detection as shown in Figure 18.6.

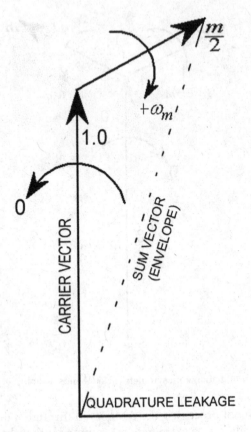

FIGURE 18.4
Asymmetrical sidebands de-spin.

The recovered baseband modulation from an asymmetric sideband situation for the envelope detector and the synchronous detector are compared in Figure 18.7.

18.3 Cross-Rail Interference

Figure 18.8 shows a typical I/Q modulator. Mixer X1 multiplies the "I" baseband signal by the cosine of the carrier (carrier at 0 degrees phase) while mixer X2 multiplies the "Q" baseband signal by the sine of the carrier (90 degrees) and sums it into combiner X5 to create the complex modulated result.

$$S = I_{BB}(\cos[\omega_c t]) + Q_{BB}(\sin[\omega_c t]) \tag{18.3}$$

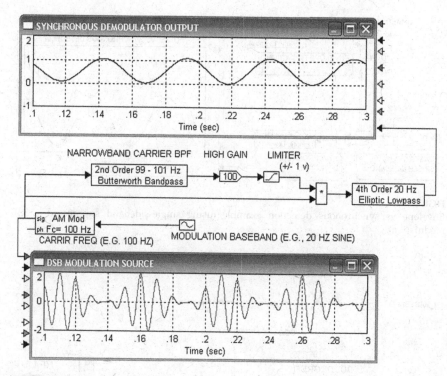

FIGURE 18.5
Double-sideband suppressed (DSB) modulation and synchronous detection.

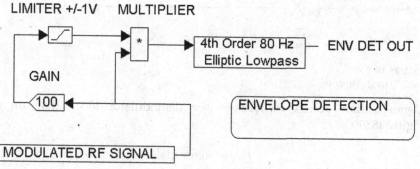

FIGURE 18.6
Envelope detection example.

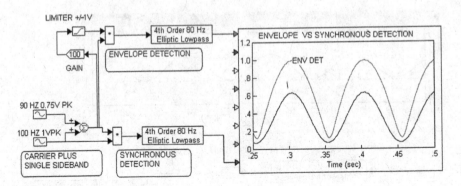

FIGURE 18.7
Envelope vs synchronous detection example (using single-sideband suppressed – SSB modulation).

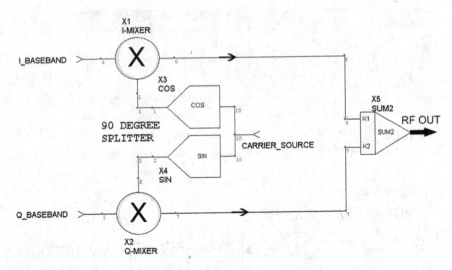

FIGURE 18.8
I and Q modulator.

Demodulation occurs as follows (see demodulator block diagram of Figure 18.9):

$$I_{OUT} = S(\cos\omega_c t) \tag{18.4}$$

$$Q_{OUT} = S(\sin\omega_c t) \tag{18.5}$$

Therefore:

$$I_{OUT} = [I_{BB}(\cos\omega_c t) + Q_{BB}(\sin\omega_c t)] \cdot (\cos\omega_c t) \tag{18.6}$$

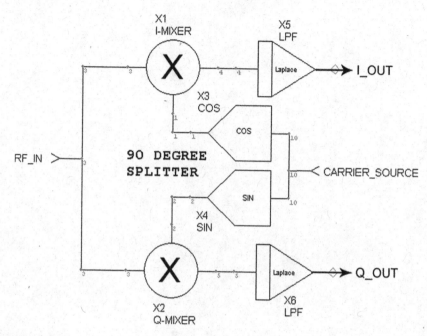

FIGURE 18.9
I and Q demodulator.

So then:

$$I_{OUT} = I_{BB}\left(\frac{1}{2} + \frac{1}{2}\cos(2\omega_c t)\right) + Q_{BB}\left(\frac{1}{2}\sin(2\omega_c t)\right) \tag{18.7}$$

and:

$$Q_{OUT} = I_{BB}\left(\frac{1}{2}\sin(2\omega_c t)\right) + Q_{BB}\left(\frac{1}{2} + \frac{1}{2}\cos(2\omega_c t)\right) \tag{18.8}$$

Thus, it is obvious that if we ignore the second harmonics at $2\omega_c$ (in a subsequent filter's stopband), we have recovered I_{BB} at a magnitude 0.5 and Q_{BB} at a magnitude of 0.5 in the ideal case.

If the filter(s) subsequent to radio frequency (RF) OUT from the modulator of Figure 18.8 has asymmetry of magnitude or delay, then signal I_{BB} leaks into the $-Q_{BB}$ recovered baseband signal and vice versa, according to our previous discussion of quadrature distortion, but this time we call it "crossrail interference." As an example, an asymmetric delay of 5 nanoseconds from band edge to band edge in a bandpass filter sandwiched between the modulator and demodulator will cause significant impairment to a 100 MBit/ second 64QAM data eye pattern if we observe the recovered baseband signal without software correction. The I and Q axes are often defined as "rails."

Bibliography

1. https://www.computerhistory.org/revolution/story/281
2. Bo Lojek, *History of Semiconductor Engineering*, Springer, 2007, ISBN 978-3-540-34257-1.
3. Sergio Franco, *Design with Operational Amplifiers and Analog Integrated Circuits*, McGraw-Hill Series in Electrical and Computer Engineering, ISBN 978-0078028168.
4. Daniel Talbot, Fiberoptic Transmission, *Journal of the Audio Engineering Society*, vol. 42, no. 5, May 1994.
5. Shiva Kumar and M. Jamal Deen, *Fiber Optic Communications: Fundamentals and Applications*, Wiley, May 2014, ISBN-13: 978–0470518670, ISBN-10: 9780470518670.
6. Eduard Säckinger, *Analysis and Design of Transimpedance Amplifiers for Optical Receivers*, Wiley, 22 September 2017, ISBN: 9781119263753.
7. H. Ugur Uyanik and Nil Tarim, Compact Low Voltage High-Q Active Inductor Suitable for RF Applications, *Analog Integrated Circuits and Signal Processing*, vol. 5, no. 3, pp. 191–194, June 2007.
8. A. Thanachayanont, et al., CMOS Floating Active Inductor and Its Applications to Bandpass Filter and Oscillator Designs, *IEE Proceedings – Circuits, Devices and Systems*, vol. 147, no. 1, p. 42, 2000.
9. J. D. Rhodes, *Theory of Electrical Filters*, Wiley, 1976, ISBN 0 471 71806 8.
10. M. E. Van Valkenburg, *Analog Filter Design*, Oxford University Press, New York, 1982, ISBN 0 19 510734 9.
11. Randall W. Rhea, *HF Filter Design and Computer Simulation*, Noble Publishing Corp., Atlanta, 1994, ISBN 1 884932 25 8.
12. Anatol Zverev, *Handbook of Filter Synthesis*, Wiley, New York, Sydney, London, 1967, ISBN 0 471 98680 1.
13. David Blackmer, *Multiplier Circuits*, U.S. Patent 3,714,462, 1973.
14. David Blackmer, *Gain Control Systems*, U.S. Patent 4,403,199, 1983.
15. Daniel B. Talbot, *Multiplier Circuit*, U.S. Patent 4,316,107, 1982.
16. Raj Senani, D. R. Bhaskar, and A. K. Singh, *Current Conveyors*, Springer, Cham, Switzerland, 2015, ISBN 978 3 319 08683 5.
17. Daniel B. Talbot, *Frequency Acquisition Techniques for Phase Locked Loops*, Wiley, 2012, ISBN 978 111 816 810 3.
18. Jacob Klapper and John T. Frankle, *Phase-Locked and Frequency Feedback Systems*, Academic Press, 1972, Library of Congress number 72-76546.
19. Daniel Talbot, Low Cost SNA Receiver Measures Across A 70-dB Range, *Microwaves & RF*, January 1997.
20. Dan Talbot, N-Over-M Frequency Synthesis, *RF Design*, pp. 54–60, September 1997.
21. Daniel B. Talbot, Ultra-High Performance Amplitude and Frequency Modulation and Demodulation, *Journal of the Audio Engineering Society*, July/August 1981.
22. Daniel B. Talbot, A Satellite Communications Broadcast Quality Amplitude Compander, *Journal of the Audio Engineering Society*, October, 1981.

23. H. Hindin, Impulse Response and Transfer Phase of Transitional Butterworth-Thompson Filters, *IEEE Transactions on Circuit Theory*, vol. 15, no. 4, December 1968.

24. Y. Peless and T. Murakami, Analysis and Synthesis of Transitional Butterworth-Thomson Filters and Bandpass Amplifiers, *RCA Review*, vol. XVIII, no. 1, March 1957.

25. C. Toumazou, F. J. Lidgey, and D. G. Haigh, editors, *Analogue IC Design: The Current-Mode Approach*, IEE, London, ISBN 0 86341 215 7.

26. Ravender Goyal, editor, *High Frequency Analog Integrated Circuit Design*, John Wiley & Sons, 1995, ISBN 0 471 53043 3.

27. Albert S. Jackson, *Analog Computation*, McGraw-Hill, New York, 1960, Library of Congress catalog card number 59-11934.

28. Robert C. Weyrick, *Fundamentals of Analog Computers*, Prentice-Hall, Inc., New York, 1969, ISBN 13: 978–0133343182.

29. J. Jess and H. W. Schussler, On the Design of Pulse-Forming Networks, *IEEE Transactions on Circuit Theory*, vol. CT-12, no. 3, September 1965.

30. R. Stephen Gordy, Simplify Acoustic Surface-Wave Designs, *Electronic Design*, 16, August 2, 1975.

31. Daniel Talbot, Go Active for Sonar or Radar Filters, *Electronic Design Magazine*, pp. 126–129, April 26, 1973.

32. Daniel Talbot, A 200 Khz to 130 Mhz Direct Frequency Synthesizer for Ultra-Low Distortion AM And FM Applications, *Proceedings of SEMICON EUROPA*, March 1981, Zurich, Switzerland.

33. Barry Gilbert, A Precision Four-Quadrant Multiplier with Subnanosecond Response, *IEEE Journal of Solid-State Circuits*, vol. SC-3, no. 4, pp. 353–365, December 1968.

34. W. R. Robins, *Phase Noise in Signal Sources*, Peter Peregrinus Ltd., London, UK, 1984, ISBN 0 86341 026 X.

35. Robert Adler, A Study of Locking Phenomena in Oscillators, *Proceedings of the I.R.E. and Waves and Electrons*, pp. 351–357, June 1946.

36. Robert D. Huntoon and Albert Weiss, *Synchronization of Oscillators*, U.S. Department of Commerce, National Bureau of Standards, Research paper RP1780, vol. 38, April 1947.

37. Behsad Razavi, A Study of Injection Locking and Pulling in Oscillators, *IEEE Journal of Solid State Circuits*, vol. 39, no. 9, pp. 1415–1424, September 2004.

38. John D'Azzo and Constantine Houpis, *Feedback Control System Analysis and Synthesis*, 2nd edition, McGraw-Hill, New York, St. Louis, San Francisco, Toronto, London, Sydney, 1966, Library of Congress Catalog Number 65-17391.

39. William F. Egan, *Frequency Synthesis by Phase Lock*, New York, Wiley 1981, ISBN 0-471-08202-3, pp. 190–191.

40. Pieter Abrie, *The Design of Impedance-Matching Networks for Radio-Frequency and Microwave Amplifiers*, Artech House, 1985, ISBN 0-89006-172-6.

41. Daniel B. Talbot, *A Review of Non-Linear Distortion Fundamentals*, Paper presented at 76th Convention, October 1984, New York, Audio Engineering Society, preprint 2134 (B-5).

42. D. Talbot, et al., *Signal Expander*, U.S. patent 4,220,929, September 2, 1980.

43. Stephen Herbert, Procedure Produces Miniature MIC Bandpass Filters, *Microwaves & RF*, April 1990.

44. Rafael M. Inigo, Gyrator Realization Using Two Operational Amplifiers, *IEEE Journal of Solid-State Circuits*, April 1971.

45. Daniel Myer, Equal-Delay Networks Match Impedances Over Wide Bandwidths, *Microwaves & RF*, April 1990.

46. W. H. Holmes, S. Gruetzmann, and W. E. Heinlein, Direct-Coupled Gyrators with Floating Ports, *Electronics Letters*, vol. 3, no. 2, February 1967.

47. Zvi Galani and George Szentirmai, DC Operation of Three-Transistor Gyrators, *IEEE Transactions on Circuit Theory*, November 1971.

48. Jochen H. W. Muller, Hybrid Integrated Gyrator in Dual-in-Line Package for Universal Application, *IEEE Transactions on Circuit Theory*, November 1971.

Index

Printed in the United States
by Baker & Taylor Publisher Services